THE GLOBE

THE
GLOBE

HOW THE EARTH BECAME ROUND

JAMES HANNAM

REAKTION BOOKS

To Alexandra and Christopher

Published by
REAKTION BOOKS LTD
Unit 32, Waterside
44–48 Wharf Road
London N1 7UX, UK
www.reaktionbooks.co.uk

First published 2023
Copyright © James Hannam 2023

Printed and bound in Great Britain
by Bell and Bain Ltd, Glasgow

A catalogue record for this book is available from the British Library

ISBN 978 1 78914 758 2

Contents

Introduction:
'The Blue Marble'

On 7 December 1972, a Saturn V rocket blasted off from Cape Canaveral, Florida. A few hours later, the Apollo 17 spacecraft escaped Earth's gravity and started its mission. This was NASA's final moon shot and the last time that anyone has left the orbit of our planet. Because the three astronauts on board happened to be flying directly into the Sun, it illuminated the full face of the Earth behind them. This hadn't happened during a manned space mission before, so all the previous photographs of the Earth had shown it partly in shadow. No one had ever seen the near-perfect disc that was receding behind Apollo 17 as it rushed towards the Moon. One of the astronauts took a picture. It has become one of the most reproduced images in history, instantly recognizable and familiar.[1]

The crew, Gene Cernan (1934–2017), Ronald Evans (1933–1990) and Harrison Schmitt (b. 1935), shared the credit for the photograph, which they called the 'Blue Marble'. It shows the outline of Africa and the Indian Ocean. As it was taken in December, the height of summer in the Southern Hemisphere, sunlight bathed the icy continent of Antarctica. Pictures of Earth from space are commonplace today and, even before 1972, impressive images had been taken by mechanical probes. But the Blue Marble is unique. It is the only occasion a human being has taken a photograph of the face of the Earth. There is an immediacy about that. No electronic wizardry intervened.

The camera, a Hasselblad with an 80mm Zeiss lens, captured the image directly onto photographic film.

In no sense was the Blue Marble necessary to prove that Earth is a sphere. But it was still meaningful to see plainly what, previously, could only be inferred. People no longer had to imagine what Earth looked like from space because someone had gone up and taken a picture. This was the apotheosis of a discovery made almost 2,500 years previously. Around 400 BC an ancient Greek had made the far-reaching suggestion that Earth might not be flat. About fifty years later, another Greek developed the theory of the Globe and showed that we live on a sphere.

Why Isn't the Earth Flat?

That the Earth is an orb may seem obvious to us. We can fly all the way around the world in little more than a day, tracking our location with our phones. It wasn't so simple a few hundred years ago. Before the sixteenth century, no one had circumnavigated the planet.

Try to forget that you were taught about the Globe at a very young age. Isn't it palpable that we live on a level plane, with some bumps and dips on account of hills and valleys? As we move around, no matter how far we travel in any direction, the world appears resolutely level. We have a strong intuition that there is an absolute direction downwards towards our feet and that this direction is the same wherever we might go. In other words, if I drop a ball in London, it feels as if its path to the ground should be exactly parallel to that of a ball dropped in New York.

How could we tell that the Earth is spherical, without the benefit of textbooks, modern technology or a theory of gravity? Someone who lives inland will often find her vision obscured by the landscape or trees. Even if she climbs as high as she can, the horizon is likely to be a distant row of hills, which in itself doesn't provide much evidence that the planet's surface is curved.

You might think that the view looking out to sea provides better evidence for the curvature of the Earth. The sharp line of the horizon results from the sea sloping away from us. If the surface of the water were flat, the interface between ocean and sky would not seem so clearly defined as they merge into each other in the far distance. As it is, standing with your toes in the surf, the horizon is just under 5 kilometres (3 mi.) away. At that distance, you can clearly see a large ship sinking below the waves as it steams away from the shore. That would be a puzzling observation if the Earth were flat. But it is even harder to understand why the sea doesn't flow off downhill. I am not trying to convince you that we really inhabit a disc, only that the planet's true shape is far from evident. Level ground corresponds to our everyday experience and there are rarely reasons to question it.

Science messes with our instincts and shows us that the world doesn't work in the way we think it ought to. It seems obvious that the Earth isn't moving. And yet, our planet is not stationary with the universe rotating around it, even though this is what appears to be happening from our vantage point. Surely if the Earth were spinning on its axis every 24 hours, we should feel it. After all, we are being carried around at about 1,600 kilometres per hour (1,000 mph) as the planet turns. If we throw something up in the air, we should not expect it to travel straight up and down. Instead, we would be carried away by the Earth's rotation so the object would fall to the ground some distance behind us. And the Earth is not just rotating; it is also travelling at the incredible speed of 108,000 kilometres per hour (67,000 mph) around the Sun.

Many of the most striking scientific theories have this counter-intuitive quality. For most of history, naturalists believed that species are fixed. The ancient Greek philosopher Aristotle (c. 384–322 BC), whom we will be hearing much more about later, asserted that the various kinds of animals had existed for all eternity. The Bible's Book of Genesis teaches that God had created living creatures, each according to their type, at the beginning of time. Since then, species had remained unchanged, inhabiting the niche that God intended for them. This is sensible given that offspring closely resemble their parents and that what differences there are, such as height or hair colour, tend to stay within strict limits. It should come as a shock that, in fact, as evolutionary theory states, we are all distant relatives of bananas (as well as to every other living organism on the planet). Nonetheless, from the late eighteenth century, the evidence of fossils indicated some kinds of animals had completely disappeared, while others seemed to have evolved into new forms over the aeons. The theory of Evolution by Natural Selection, first developed by Charles Darwin (1809–1882), explained how species can change and provided the mechanism by which they do so.

A few decades later, Albert Einstein's (1879–1955) theory of Special Relativity showed that time itself moves more slowly as objects approach the speed of light, and that mass and energy are equivalent. Meanwhile, a host of physicists developed quantum mechanics, which revealed a subatomic world where probabilities replace the certainties of Newtonian physics.

Successful theories, like Relativity and Evolution, may be counter-intuitive but they still need to account for how we experience everyday life. Einstein's formulas might reveal time can dilate, but they also show that this phenomenon disappears at the low speeds we are used to. Evolution acts over long periods of time, so its effects cannot be seen in a few generations. And the idea that Earth orbits the Sun only became accepted once it was understood why we cannot feel it move. The same is true for the theory of the Globe. Anyone who is going to seriously entertain that we live on a ball is going to have to explain why it appears to be flat, and why people on the other side don't fall upwards into space.

Regardless of whether anyone knew, it's always been an objective fact that the Earth is spherical. However, it was not always reasonable to believe it. We will meet many intelligent individuals in these pages who didn't know the true form of our planet. They were not fools – something we need to bear in mind throughout this book. Nineteenth-century Europeans were quick to label people living on other continents, or from the past, as ignorant because they didn't grasp that the Earth is round. Unfortunately, you still hear echoes of that attitude today. This has led some historians to anachronistically search for evidence for the Globe (especially in holy books) where it is not to be found. People who lived before us do not need to be defended from the accusation that they held views, such as believing the world is a disc, that they really did hold. But neither should we allow anyone to cast aspersions upon them for not being aware of what we happen to know now.

The discovery that the Earth is a sphere was not a realization of the obvious. It was humanity's first successful scientific theory. Some innovations, like agriculture, arose independently on more than one occasion. Archaeologists believe that farming appeared in at least three places: China, the Middle East and Mesoamerica. Evolution by Natural Selection was independently devised by Alfred Russel Wallace (1823–1913) and Charles Darwin. However, as far as we can tell, the shape of the Earth has been discovered just once, in ancient Greece, and took over 2,000 years to spread from that point to the rest of the world. The fact the theory of the Globe does not appear to have arisen anywhere else gives us a strong indication of just how divorced from common sense it is.

The flat-earthers weren't always who we might expect. For example, there's a widespread calumny that medieval Europeans thought the world was shaped like a dinner plate. The truth is that, after AD 800, we don't know of anyone in western Europe with a modicum of literacy who didn't think that the Earth is spherical. Indeed, science in the Middle Ages was a vibrant tradition, which laid important foundations for the discoveries of later researchers like Galileo Galilei (1564–1642).

Meanwhile, the tall tale that Christopher Columbus (1451–1506) proved the Earth is round has entered the bloodstream of popular culture. In the 1937 film *Shall We Dance*, Ginger Rogers sang Ira Gershwin's lyric about how everyone laughed at Christopher Columbus because he said the world was round. The song 'They All Laughed' has since been recorded at least twenty times, including by Bing Crosby, Frank Sinatra and Lady Gaga. In March 2012 President Barack Obama repeated the myth in a speech about renewable energy. He compared those who are sceptical about low-carbon electricity generation to the alleged opponents of Columbus's voyage. 'If some of these folks were around when Columbus set sail, they must have been founding members of the Flat Earth Society,' he said. 'They would not have believed that the world was round.'[2]

The fact is, by 1492 the theory of the Globe wasn't controversial in Europe. The royal advisers, who thought Columbus sailing off into the vast emptiness of the Atlantic was a daft plan, didn't doubt that eventually he'd end up in India, just that there was no way he'd survive the journey. Nevertheless, once upon a time, everyone assumed the world was flat. Many continued to do so right up until the twentieth century. A few still do.

This book describes how we came to know it isn't.

1

Babylon: 'The four quarters of the Earth'

Some of the earliest clues about how ancient people saw their world are found in the Babylonian *Epic of Gilgamesh*. It recounts the quest of Gilgamesh, king of Uruk, who sought the secret of immortality beyond the edge of the world. His best friend had died, leaving him heartbroken, and showing that death was the one enemy he could not defeat. Leaving his home city, he trekked through a cedar forest until he reached Mount Mashu, behind which the Sun hid at night. Although climbing the mountain was impossible, there was a passage through its bowels guarded by a monstrous beast, half-man, half-scorpion. Following some negotiation, the creature permitted Gilgamesh to enter the tunnel. He crept through complete darkness for several days before emerging into the Garden of the Gods, where the Sun sojourned between dusk and dawn. Beyond the garden was a sea. The hero took a boat to the far side and met a man called Utnapishtim. The gods had revealed to him the magic plant that was the secret of immortality. Gilgamesh acquired some of its life-giving leaves but, on his way home, a serpent stole them. Thus he returned to Uruk empty-handed.

Gilgamesh and Cuneiform

The ancient Greeks called the area where Gilgamesh lived Mesopotamia, meaning 'the land between the rivers'. The rivers in question are the Tigris and the Euphrates, which have their sources, a mere 80 kilometres (50 mi.) apart, in the Tarsus Mountains of eastern Turkey, before flowing through modern Iraq. On their journey to the Persian Gulf, they sometimes come very close to each other, but their confluence today is just 190 kilometres (120 mi.) from the sea. The Tigris and Euphrates supplied the water for an irrigation system that supported one of the earliest urban societies. By 3000 BC, mud-brick cities stood on their banks, surrounded by agricultural hinterlands. Inevitably, centuries of intensive farming led to salt accumulating in the fields, returning them to desert and causing some of the cities they sustained to be abandoned. Other settlements were razed as empires rose and fell. Today, many are simply forgotten mounds, called tells, waiting for archaeologists to dig them up.

Gilgamesh's Uruk was a real city in southern Mesopotamia, founded in the fourth millennium BC. Gilgamesh himself is recorded in a list of its kings in the Sumerian language, the very earliest known written script, so he is generally accepted to be a real person who lived around 2700 BC.[1] There are references to him in hymns sung within a few centuries of his death, although the full epic as we know it was not composed for at least another thousand years.

There's no congruous description of the world in the story of Gilgamesh's search for immortality. Perhaps the cedar forest stood for the wooded slopes of Lebanon to the west, while the Garden of the Gods was more likely to signify fertile land down by the Persian Gulf. As for Mount Mashu, it is usually located in the far north and represented the boundary between the lands of men and the realm of the gods. Mashu performed the important function of explaining why the Sun was not visible at night. In any case, despite his confusing geography, the poet who composed the *Epic of Gilgamesh* assumed the Earth was flat.

Relief depicting
Ashurbanipal hunting
lions, North Palace of
Nineveh (Iraq),
c. 645–635 BC.

We owe much of our knowledge about Gilgamesh and
other figures in Mesopotamian literature to the later empire of
Assyria. It was the most rapacious of ancient powers, yet some
of its brutal kings were also proud of their literacy and respect
for the past. The Assyrian conqueror Ashurbanipal (d. 631 BC)
liked to be depicted with writing implements tucked into his belt,
even while he was smiting his enemies. He sent agents across
his empire to obtain texts for the magnificent library in his capital of Nineveh. At the library, professional scribes copied the
looted literature using a cuneiform script, formed from wedges
impressed into wet clay by a stylus.[2] The scribal vocation was
of high status and ran in families. Thanks to the patronage of
Ashurbanipal and the skill of the royal scriveners, his library

contained a remarkable range of documents, including the *Epic of Gilgamesh*, written in lucid cuneiform.

In about 612 BC, Nineveh was pillaged and burned. At the time, it was the greatest city in the world, but its devastation in a rebellion against the predatory rule of its kings brought their empire crashing down. The library was at the centre of the conflagration. Fire is the mortal enemy of scrolls and books. But the writings in Nineveh were inscribed on thousands of clay tablets. When the city burned, the tablets were baked hard by the intense heat, making the flames an agent of their preservation.[3]

Today, the British Museum in London holds about 130,000 tablets from the library of Ashurbanipal and elsewhere.[4] Perhaps a million more have been excavated in the Middle East and are stored in museums and universities. When scholars first deciphered the British Museum's tablets in the nineteenth century, their findings caused a sensation. In particular, they found fragments of stories, such as one about a great flood, which seemed to be precursors of tales from the biblical Book of Genesis. Such was the public excitement that, in 1873, the *Daily Telegraph* financed an expedition to Iraq to find more tablets.[5]

We now know the flood story is part of the *Epic of Gilgamesh*. It was recounted by Utnapishtim, an immortal man who had survived the deluge because a god called Ea instructed him to tear down his house and build an ark from the materials. The flood began when the gods stirred up a great storm, causing the waters to overwhelm dry land. The occupants of the ark, alone, survived. After seven days, the waters started to recede and Utnapishtim released a dove, a swallow and a raven to see if there was any dry land where he could disembark. The ship finally alighted on Mount Nisir in northern Iraq, some way south of Mount Ararat, where Noah's Ark ran aground. Utnapishtim sacrificed a sheep to the gods, who rewarded him and his wife with immortality.[6]

The Babylonian Creation Myth

Prior to the middle of the second millennium BC, the Sumerians ruled much of southern Mesopotamia. They were the ones who developed cuneiform writing, recorded the deeds of Gilgamesh and worshipped the pantheon that provided the supporting cast in later Mesopotamian mythology. Their kings called themselves the rulers of the four quarters of the Earth, or kings of the universe, to show that their realm encompassed the whole world.

As the power of Sumer faded, other empires rose in its place and adopted its nomenclature. Hammurabi (*c.* 1810–*c.* 1750 BC), the first great monarch from Babylon, styled himself 'King of Sumer and Akkad, King of the Four Quarters of the World'. At least he stopped short of declaring he was king of the universe.[7] Later still, Assyrian monarchs such as Ashurbanipal used the same honorific, although he may have been required to prosecute military campaigns to the north, south, east and west before he could legitimately do so.

In this context, the words 'four quarters' can be translated as 'four corners', implying a square or rectangular shape, but it's probable that the corners refer to the internal angles of four quadrants of a circle joined at the centre of the world. In some texts, the centre is a mountain or the source of four rivers. In others, it might be the city of Babylon or, more precisely, the temple at its core.

The phrase 'the four corners of the Earth' has been a familiar idiom in English since long before the decipherment of cuneiform because it entered the language from the Bible. In about 750 BC, the prophet Isaiah predicted that God would 'gather together the dispersed of Judah from the four corners of the Earth' (Isaiah 11:12). Isaiah used the phrase to assert the power of his god over the entire world, but he might have been deliberately echoing the pretensions of the Assyrian kings, who were encroaching upon the Kingdom of Judah at the time.

Among the many texts collected by the Assyrians and preserved on the cuneiform tablets in the British Museum is a Babylonian creation myth. It's known as *Enuma Elish* after its first two words, translatable into English as 'When on high'. Dating from early in the first millennium BC, *Enuma Elish* recounts how Marduk, Babylon's divine patron, created the world and became the supreme deity. It was recited in full during the new-year rites each spring. During this week-long festival, the gods of all the cities dominated by Babylon, represented by their cult statues, visited the great temple of Marduk to pay homage to him as their king.[8] This justified the position of Babylon as the leading metropolis of Mesopotamia, owed loyalty by other cities, just as Marduk expected fealty from his fellow immortals.

According to *Enuma Elish*, the universe arose from a mixture of salt and fresh waters. Later, the gods fought a series of wars, before Marduk earned a final victory over his ancestor, a monster called Tiamat, who represented salt water. Marduk slew her and, after the battle, created the heavens and Earth from her corpse: 'The lord [Marduk] rested; he gazed at the huge body, pondering how to use it, what to create from the dead carcass. He split it apart like a cockleshell, with the upper half he constructed the arc of the sky.'[9]

Tiamat's breasts became mountains, while the Tigris and Euphrates flowed from her dead eyes. Marduk erected the gates of the dawn and the dusk from her ribs to the east and west, appointing the sun and moon gods to travel across the sky by day and by night. Then, the god Ea (who would later warn Utnapishtim about the flood) fashioned humanity from the blood of one of Tiamat's allies. At Marduk's instigation, the gods immediately indentured the humans as their servants, slaves by their very nature.

The universe portrayed by *Enuma Elish* comprised a series of layers. The world of the living was in the middle, formed from the remnants of Tiamat. Just beneath was an abyss filled with sweet fresh water. The hide of Tiamat formed a lid that kept the roiling deep in check. Beneath the abyss was the subterranean realm of the dead. Mesopotamian literature portrayed hell not as a fiery furnace but as a wasteland like the desert that surrounded the fields irrigated by the Tigris and Euphrates. The sky was a dome or vault that covered the Earth, with the gods living in the highest heaven above this firmament.[10] According to one source, the heavens had three tiers lined with red, green and blue stone respectively.[11]

The great ziggurats of Mesopotamia bridged the gap between heaven and earth, supporting stairways down which the gods could descend. Next to each ziggurat stood the main temple of the city, thought to be the fulcrum of the world. There, its divine patron was mollycoddled by the priesthood through ritual feeding of its idol. In return, the god attended to the important business of maintaining the structural integrity of the universe and keeping chaos at bay.[12]

The world picture of the Babylonians is illustrated by a map from the sixth century BC, preserved on a clay tablet in the collections of the British Museum. It shows Babylon on the Euphrates near the centre, surrounded by neighbouring cities and kingdoms. The known world is circular and ringed by an ocean of bitter salt water with a range of enormous mountains, described as a wall, to the north. On his journey, Gilgamesh was able to travel past this mountainous boundary to places that only gods and heroes can go. While the map is figurative, there is no reason to doubt it reflects how Babylonians saw the Earth as a whole – a disc girted by sea and mountains.[13]

Babylonian map of the world, 6th century BC, clay tablet.

Babylonian Astronomy

The Babylonians were assiduous astrologers, driven to observe the heavens by a desire to know the future. They took it for granted that celestial phenomena were messages from the gods that told them something about forthcoming events. Obviously, the gods would not use the heavens to relay messages to common folk, so events like eclipses had to be warnings intended for the king about matters of national importance. To ensure that no divine messages were missed, a class of professionals was responsible for watching out for omens. The assumption that the sky was a medium for the immortals to communicate with Babylonian or Assyrian monarchs implies a rather parochial view of the universe, but this was of a piece with believing the king was the rightful ruler of the entire world.

As far as Babylonian astrologers were concerned, any celestial event could be laden with meaning. Eclipses were almost always bad news. An eclipse of the Moon warned of death or defeat for the king. The part of the lunar disc that was occulted indicated from which of the four quarters of the world disaster might strike. Thankfully, if cloudy weather meant that the Moon wasn't visible where the king was staying, he could breathe easy.[14] Otherwise, the priests would recommend steps to protect him from the omen. He might have to engage in ritual washing, or confine himself to a special house for a few days. Serious cases called for more severe action. The monarch temporarily abdicated and his place was taken by a prisoner. The wretch found himself sat on a throne and treated with the utmost respect by the highest men in the land. After a couple of weeks, when the danger of the eclipse had passed, the real king resumed his place. Meanwhile, the prisoner was taken away and executed.[15] The gods, after all, must have their due. This ritual didn't always go to plan. In around 2000 BC, a king called Irra-imitti was warned of terrible portents that foretold his death. To protect him, the priests shut the king in his palace and installed a gardener on the throne. This was supposed to be temporary, but Irra-imitti fatally choked on his porridge during his confinement. Following this clear expression of divine will, the lucky gardener, who had expected to be killed, found himself acclaimed as king permanently and enjoyed a successful reign of over twenty years.[16]

After a couple of millennia of careful observation, Mesopotamian astronomers spotted some regularities in the heavens. Already, by 2000 BC, they had realized that the morning and evening stars, the brightest lights in the night sky after the Moon, were the same object. The Sumerians identified it with Inanna, goddess of love and war, and today we call it Venus.[17] Because Venus orbits nearer to the Sun than Earth does, it always sticks close to the Sun when visible in the sky, appearing shortly before dawn as the morning star or after dusk as the evening star. Babylonians also knew of four other planets, which we call Mercury, Mars, Jupiter and Saturn. They noted that the planets periodically disappeared below the horizon and believed them to reside in the underworld.

Babylonian astronomers recorded eclipses incessantly and came to recognize that they could only occur at certain times: at a full moon for a lunar eclipse and at a new moon for an eclipse of the Sun. Over time, they collected enough data to determine that the frequency of these portents follows long-term patterns and that it was possible to guess when they were more likely. The most useful of these criteria was a period of eighteen years, that is, 223 lunar months, nowadays called the Saros cycle. At the end of this period, the Moon returns to the position relative to the Sun where it had started, so the sequence of eclipses resets.[18] Nonetheless, the Babylonians never figured out why eclipses happen. They were messages from the gods, not physical facts or events to be explained.[19] Nor did they grasp that the light of the Moon was reflected from the Sun.[20] And because they thought that Earth was a disc, they imagined that, when the stars and planets fell below the horizon, they sank beneath the ground. Their sophisticated mathematical models of the heavens were calculating devices and not reflections of reality.[21]

The Mesopotamian assumption that the Earth was a disc found its way into their myths, the titles of their kings and the surviving example of one of their maps. As the Tigris and Euphrates rose and fell, it made sense to imagine that the ground was like a hide stretched over primeval waters that could burst forth as devastating floods. More generally, Babylonians saw themselves living at the centre of the Earth, surrounded by the known world, which was bounded by impenetrable mountains or a trackless sea. Above and below, heavens and hells took their shape from myths like *Enuma Elish* that were themselves composed for polemical purposes. Thus the Babylonian world picture reflected the geopolitical circumstances of Mesopotamia. It was a vision informed by the particular circumstances of the people who developed it. A civilization that arose in a different environment would not see the world in the same way. Egypt was a case in point.

2

Egypt: 'The black loam and the red sand'

Egypt, wrote the ancient Greek historian Herodotus (*c.* 484–*c.* 430 BC), is the 'gift of the river'.[1] Satellite images show us just how true that is. From space, the thin emerald snake of fertile land on either side of the Nile, threading its way through the vast desert, appears fragile and ephemeral. Yet, the Nile has maintained civilization in Egypt for thousands of years. The desert protected the Egyptians from external invaders, while the river encouraged internal cohesion. It was continually navigable from the delta in the north for over 1,000 kilometres (600 mi.) to Aswan in the south, where the first cataract intervened. Egypt, united under a single pharaoh before 3000 BC, retained its integrity and independence for much of the next 2,500 years.

Each summer, the Nile would surge, carrying water and rich sediment across the fields laid out on each bank. The Egyptians devised a system of dykes and ditches to spread the fecund waters over as much farmland as possible. When the flood ebbed, peasants could plant their crops in the mud. By ancient standards, the Nile valley was incredibly productive. Common threats to pre-modern agriculture, such as salination and soil erosion, did not arise as the floodwaters refreshed the fields each year. That allowed Egypt to support a population of between 2 and 4 million well before 1000 BC.

The Primeval Mound

In the beginning, said the Egyptians, there had been nothing but churning chaos. One day, a little hillock appeared in the midst of the deep. It's easy to imagine that when they watched the floodwaters of the Nile recede, people would have seen how slightly higher ground emerged first and rapidly filled with life. This was, perhaps, the inspiration for the primeval knoll with which the world began. Several temples claimed to preserve the original mound and encouraged pilgrims to view it.[2]

Accounts differ about how the gods were born. According to a dominant tradition, the supreme deity, Atum, found himself alone on the mound and so produced two children from his semen to keep him company. The children, Shu and Tevnut, had a pair of offspring of their own, Geb and Nut. These two married, but ended up having to be separated after they had a fight. Nut had been eating their children, the stars, and Geb was not happy about it. Nut arched herself over Geb and became the sky, while her brother lay beneath her as the Earth. Together, they formed the boundary between the realm of order within and the chaotic waters seething without. Nut's children, the stars, are still visible within her body during the night.[3]

The Sun was personified by the god Ra. According to a diagram on the ceiling of the tomb of the New Kingdom pharaoh Seti I (1323–1279 BC), Ra crossed the sky from east to west each day in his boat (boats being the accepted mode of royal transport in ancient Egypt). Ra's origins are recondite. One account says he was born from a blue lotus flower growing in the primordial ocean, another that he hatched from an egg. In any case, he was an important deity from an early date, served by his priesthood in the city of Heliopolis, near the great pyramids. Every evening, when he had completed his journey across the sky, Ra needed to return to the eastern horizon so he could

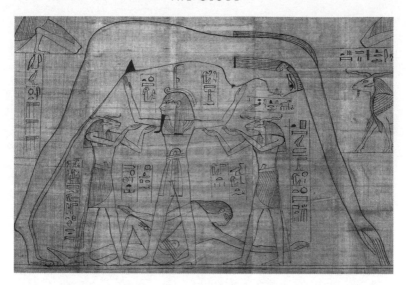

Geb and Nut, detail from the Greenfield Papyrus, Thebes, Upper Egypt, c. 950–930 BC.

rise the next day. The pictures in the tomb of Seti I suggest he made his way back during the night, perhaps travelling within the body of Nut so that his light was not visible. At dawn he emerged, rejuvenated and ready to take to his boat once again.[4]

Other accounts, such as the *Book of the Caverns* from the late second millennium BC, tell a more detailed story. Ra sailed across the sky each day in one boat, complete with a hard-working crew of minor deities. Then, at dusk, he boarded another vessel to voyage through the underworld, finishing the journey at sunrise. He was joined by an armada of ships that helped him to defeat the serpent Apophis, which he had to slay before the new day could begin. Every night this great snake attacked, and every night Ra prevailed so that the Sun could reappear.[5] In an intriguing detail, we are told that the length of Ra's underground journey was 24,000 kilometres (15,000 mi.).[6] While it is tempting to treat this as an indication of how wide the Egyptians thought that the world was, it is more likely to mean 'a long way'.

The Egyptian Universe

Despite disparate stories about the gods and the path of the Sun, the ancient Egyptian world picture was reasonably consistent. The Earth was a flat disc over the waters of the abyss, from which the Nile had its source. The sky was a vaulted roof, held up by four pillars or mountains, which kept out the waters from above, making the world a bubble of order within an endless sea of mayhem. The hieroglyph for the sky resembles a shallow arch while the solid columns at the Temple of Karnak stood comparison to the supports of heaven itself.[7]

To the east and west, the horizons represented the edges of the world. Such was their symbolic importance, they were governed by a specific god, Horakhty, an aspect of the falcon-headed Horus. The Nile flowed down the middle of the Earth, with its source at the top and the Mediterranean Sea at the bottom. Egyptians sharply distinguished between the thread of fertile black soil on each bank of the river and the red sand of the desert on each side. They also demarcated the marshy delta by the sea, and the Nile valley above it. Throughout its history, the southern part of the country was called Upper Egypt and the Nile delta Lower Egypt. The pharaoh, with his double crown, was king of the two lands.

Egypt was a conservative society, centred around the person of the pharaoh and with a deep suspicion of foreign cultures. Their world picture reflected these attitudes. It distinguished between the land of order, being Egypt under the governance of the pharaoh, and the realm of chaos, which was everywhere else. The further Egyptians travelled from the Nile, the closer they got to the ultimate horizon, beyond which the primeval waters threatened unbridled havoc. Only the enclosing barriers of the sky and the Earth kept the deluge at bay. For the Egyptians as much as the Babylonians, geography was imbued with meaning, including the centrality of their civilization in the design of the universe.[8]

There's a remarkable illustration of the Egyptian world picture on the granite sarcophagus of a priest called Wereshnefer, who died around 350 BC. He was buried in the necropolis of Saqqara, south of Cairo, although his enormous stone coffin is now in the Metropolitan Museum of Art in New York. It shows the Earth as a disc of concentric circles. Nut arches her body over the whole, while Geb supports it from below. Hostile tribes menace civilization from the sides, while, at the top, the Nile issues forth from an underground cavern. Egypt itself is represented by a ring of the symbols of its forty constituent districts, while sky and stars appear within a circle at the centre. Even though this image dates from near the end of ancient Egyptian civilization, it would have been recognizable to Wereshnefer's predecessors in the priesthood for two millennia previously.

Despite its long-lasting and highly developed culture, Egypt's scientific achievement was modest. The priests of Ra, responsible for watching the sky, were aware of some constellations as well as the five planets visible to the naked eye. But there is no evidence that they made any systematic observations of the planets' wanderings, nor could they predict eclipses.[9] Lacking a tradition in astrology, they had no incentive to do so. Determining the calendar was the main impetus for studying the heavens. The path of the Sun provided the basis for keeping track of time, which led the Egyptians to follow a 365-day year. Priests also recorded the stars visible in the sky to help them mark the seasons and especially when the Nile should start to flood. Later, the Romans adopted the Egyptian's 365-day year and, albeit with a few tweaks, we still use it today.[10]

The ancient Greeks had an obsessive regard for the wisdom of Egypt's priests and credited them with all kinds of esoteric knowledge. Early Greek sages, such as Thales (*c.* 624–*c.* 548 BC)

Depiction of the world, from the sarcophagus of Wereshnefer,
Saqqara, Egypt, 380–300 BC.

and Pythagoras (*c.* 570–*c.* 490 BC), whom we will meet in the
next few chapters, were said to have acquired their perspicu-
ity during visits to the land of the pharaohs. The philosopher
Plato (*c.* 428–*c.* 348 BC) claimed that tales about the lost city of
Atlantis came from the same source. Sadly, when hieroglyph-
ics were finally deciphered in the early nineteenth century, any
expectations that Egyptian literature would reveal the origins

of philosophy were disappointed. While papyrus fragments recovered from the sands preserve impressive practical mathematics, they show little interest in astronomy and certainly nothing about Atlantis.

Though the civilizations of Mesopotamia and Egypt were in contact through trade and the occasional war for thousands of years, it is striking that neither made any space for the other in their world pictures. They each saw themselves as the focal point of the universe. In 600 BC, it seemed these twin pillars of urbanized society were as strong as ever. Egypt was on its 26th dynasty of pharaohs, while Mesopotamia was ruled from Babylon by Nebuchadnezzar (d. 562 BC), who would shortly add Judah and Jerusalem to his domain. But to the east and west, outsiders were stirring who would sweep away the old order and fight for supremacy for the next thousand years. These cultures, the Persians and Greeks, would foster new models of the universe that were unlike those that had gone before.

3

Persia:
'Order and deceit'

For thousands of years, the steppe that runs from the borders of China to the Hungarian plain has been Eurasia's great throughfare. The boundary between the land and the sky was a line of low hills that surrounded the viewer in all directions. Beyond those hills, there was another range just like it. That was all there was to see, except for the rivers that cut across the grassland. Tribes moved with the seasons as they followed their herds, but rarely travelled outside the steppe itself.

It was a tough life. But, from about 5000 BC, a group of nomads from north of the Black Sea, around the area of modern Ukraine, achieved a series of breakthroughs that transformed their existence. Perhaps inspired by the example of early farmers encroaching onto the edges of the plains, they stopped treating the local horses as wild animals to be hunted for food and domesticated them. They also invented or heard about the wheel. Combining the horse and wheel produced another innovation: the wagon. And they developed a lighter vehicle with two spoked wheels for when they wanted to move rapidly: the chariot.[1] This had obvious military applications that they were not slow to exploit. Around 3000 BC, the nomads started to spread in all directions. The group that went west had subjugated Europe within a thousand years. Others wended east and south. Everywhere they went, they imposed their culture on the indigenous farmers. As a result, modern languages ranging from English and French to Persian and Hindi have a common root in the tongue spoken by these tribes. This language family is called Indo-European.

One of the tribes crossed the Hindu Kush and entered India before 1000 BC. Another group went west, founding the Kingdom of the Mitanni in northern Syria. It's here that archaeologists have found the earliest inscriptions in an Indo-European language. Yet another branch moved south, crossed the Zagros Mountains and settled in an area called Elam. These people were called Pars, from which the ancient Greeks derived their moniker for them – the Persians. However, in their earliest traditions, they called themselves 'Aryans' or, as it has been spelt for much of their history, 'Iranians'.

Zoroaster and the Struggle between Order and Deceit

Even before they settled in Elam, the ancestors of the Persians had composed a series of religious hymns called the Avesta. The dialect in which the Avesta is recited and the places mentioned in it suggest the hymns were first performed from the middle of the second millennium BC, in Afghanistan and northern Iran. After being passed down orally for many generations, they were finally written down in the third century AD.[2] They are the foundational texts of Zoroastrianism, and Zoroaster himself is mentioned many times within. Whether he was a real person, let alone the prophet who founded the faith named after him, is a contentious question. Many scholars do believe a historical figure lies behind the name.[3]

The Avesta presupposes a great battle between 'Order' and 'Deceit', personified by the supreme god Ahura Mazda and the demon Ahriman respectively. The religious duty of Zoroastrians is to carry out the rituals and sacrifices necessary to aid Ahura

Mazda in his struggle with the demon, helping to ensure his final victory and the restoration of perfection.

The fragmentary nature of the sources makes it difficult to reconstruct the early Avestan world picture. Many of the extant texts were compiled by desperate Zoroastrians trying to salvage something of their traditions after the Arab invasion in the seventh century AD. According to surviving pieces of the Zoroastrian scriptures, the Earth is a disc held up by mountains in the centre of the sky. These form a circuit of 2,244 peaks around its edge.[4] Since the Persians and related tribes had been wanderers who lived inland, we would expect them to imagine the world was circumscribed by a mountain range rather than an ocean. At the centre of the Earth was Mount

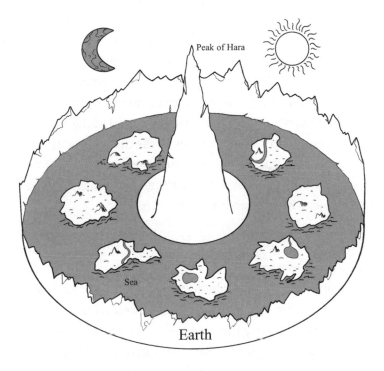

The world according to Zoroastrian cosmogony, after Mary Boyce, ed. and trans., *Textual Sources for the Study of Zoroastrianism* (1984).

Hara, the axis of the universe, around which the Sun, Moon and stars revolved. At the beginning of time, Zoroastrians believed that Ahura Mazda had created a world so flawless that it didn't move (perhaps because any change could only render it less perfect). However, in the conflict with the diabolical Ahriman, the world was damaged and began to rotate about Mount Hara, giving us night and day.[5] Rain fell, and the waters formed seas that divided dry land into seven continents.[6]

There are not many commonalities among the Babylonian, Egyptian and Zoroastrian world pictures. True, all three civilizations assumed the Earth was flat, that the heavens above were a home of the gods and that the underworld was a place of monsters and darkness. Beyond that, the way they imagined the Earth depended on their local environment. For the Egyptians, it was bisected by the Nile and bounded to the east and west by the far horizons. The Babylonians said it was like a drum skin spread over the waters of chaos, with their capital city in the middle. Both civilizations thought that water was the primordial substance, since it was so foundational to their agrarian societies. In contrast, Zoroastrians surrounded the Earth with the ring of highlands that they saw in the distance, with Mount Hara at its hub. As we will see, the motif of a central massif would later appear in writings attributed to another branch of the Aryans, who migrated to India.

The Persian Empire

By the time the Persians had formed their own kingdom in the sixth century BC, it's likely that Zoroastrianism was the dominant faith among the ruling aristocracy. The monumental inscriptions of their kings primarily celebrate military victories, but these carvings also attribute their royal office to Ahura Mazda.[7]

The Cyrus cylinder: Babylonian account of the conquest of Babylon by
Cyrus in 539 BC, after 539 BC, clay.

The Persian Empire began in the reign of Cyrus the Great
(600–530 BC), who ascended the Persian throne in 559 BC. He
launched a lightning campaign of conquest that took over the
territory of the Assyrians as well as the rest of Asia Minor. Then
he invaded Mesopotamia, capturing Babylon in 539 BC. He went
out of his way to respect the traditions of the Babylonians. On a
clay cylinder covered in cuneiform script, which he buried in the
foundations of the Temple of Marduk, Cyrus took the ancient
title 'King of the Four Quarters of the World' and boasted how
he had restored the ancestral cults of Babylon.[8] In the following
fifty years, the Persians also conquered Egypt to the west and
the Indus Valley to the east. Their forbearance of the traditions
of the populations they conquered allowed them to build and
hang onto the largest empire the world had ever seen.

Despite Cyrus's show of respect, the Persian invasion
marked the end of Babylon's hegemony over Mesopotamia.
Even when the area had been ruled by the Assyrians, Babylon
had remained the cultural lodestone in the region. But now the

old rites slowly fell into decay. The Persians had little interest in Mesopotamian religion, even if they tolerated it.

Babylon's astrologers found themselves in a quandary. They could no longer rely on the patronage of kings keen to hear about messages from heaven, so they needed to find a new market. In the fifth century BC, they began to provide horoscopes to private individuals, making predictions about the course of someone's life from the position of the planets at the moment of their birth.[9] This necessitated some systemization in the heavens. By 400 BC, astrologers had carved up the path of the Sun across the background of the stars into twelve equal sectors each named after a prominent constellation within it. They divided the annual circuit of the Sun into 360 degrees, so it

Modern zodiac chart incorporating late 15th-century woodcut zodiacal signs.

moved about one degree a day. The twelve houses of the zodiac remain the basis for horoscopes today.[10] And, we still split circles into 360 degrees, following the Babylonian number system, which they based on sixty rather than ten. That's why we also have sixty seconds in a minute and sixty minutes in an hour.

The new system of astrology became immensely popular. No longer was knowledge of the future only for kings with whom the gods deigned to communicate. Now, a birth chart was available to anyone who could afford the fee. Horoscopy spread to the Greek and Roman world, to India and, later, even to China. Along the way, it was the main driving force for the enhancement of astronomy. Astrologers needed to know the past and future positions of the planets against the backdrop of the zodiac. The only practical way to do that was to develop mathematical formulas that could predict the movements of the stars.

Persia versus Greece

By 500 BC, the Persian Empire had embraced the Aegean coast of Asia Minor, a region called Ionia, where Greek immigrants had founded several colonies. After these Ionian cities rebelled in 499 BC, it took six years for Persia to regain control. The epicentre of the insurrection was the city of Miletus, whose population the Persians enslaved after it fell in 494 BC. Blaming the seriousness of the revolt on the support that the Ionians received from mainland Greece, the Persian king, Darius I (550–486 BC), decided to expand his realm across the Hellespont into Europe. This didn't quite go to plan. His army managed to overpower Macedonia and Thrace, to the north of Greece, but suffered a catastrophic defeat to Athens at the Battle of Marathon.

Ten years later, Darius's son, Xerxes (518–465 BC), tried again. This time, the Persians took no chances and assembled an enormous army and fleet. The Greeks' position looked hopeless.

Mosaic depicting Alexander the Great, Pompeii, c. 100 BC.

They faced a united empire that could field troops from as far afield as India. The Greeks had plenty of practice fighting each other, but found it hard to unite their forces. Nonetheless, things went badly for Xerxes. The Athenians' galleys destroyed his fleet at the Battle of Salamis, while a Greek coalition led by Sparta crushed his land forces at Plataea. The war dragged on for decades, but in the end the Greeks blunted the Persian threat.

In the 330s BC, the struggle with Persia resumed. By then, the fractious Greeks had been unified under the yoke of Philip II (382–336 BC), the king of Macedon. The citizens of Athens considered Macedonia a militaristic backwater but could not resist its armies. After Philip's assassination, his son, Alexander (356–323 BC), led an expedition that he claimed was to take revenge on Persia for the wars over a hundred years before. At Gaugamela in northern Iraq, Alexander's Macedonians were thousands of miles from home and faced 100,000 troops led by Darius III (381–330 BC). By all accounts, Alexander should have lost. Instead, his victory destroyed the Persian Empire. He marched on Babylon and then to Darius's ceremonial capital, Persepolis, which he burned to the ground. By the time Alexander died, he had conquered Asia Minor, the Levant, Egypt and Persia itself, not stopping until he reached India. Along the way, he had founded numerous cities, many of which he modestly named Alexandria.

The conquests of Alexander the Great spread Greek culture all the way to Afghanistan but his empire didn't outlive him. On his death in Babylon, his squabbling lieutenants dismembered his vast domain. One of them, Ptolemy (367–283 BC), took his body to Alexandria in Egypt where he founded a new dynasty of pharaohs. Ptolemy made Alexandria into a font of Hellenistic civilization, and it remained an important intellectual centre until the seventh century AD.

Despite the depredations of Alexander, the Persians would come roaring back. A new Zoroastrian dynasty, the Sassanids, arose in the second century AD and reconquered much of the old Persian Empire. We'll meet the Sassanian Persians later, but by then, we'll have seen how it was the Greeks who reordered the cosmos and became the first people to abandon the flat earth.

4

Archaic Greece:
'The Shield of Achilles'

The most ancient of the surviving Greek descriptions of the universe isn't from a religious or scientific source but an epic poem about the Trojan War. Homer's *Iliad*, conventionally dated to the eighth century BC, tells of how a sulk by the hero Achilles set off a spiral of violence. He had been denied the enjoyment of a captive girl by his king, Agamemnon, leader of the Greeks, and so was moping around in his tent instead of fighting. However, Patroclus, his best friend, wanted to go out and earn some glory for himself. He borrowed Achilles' armour and went off looking for battle, which he found at the hands of Hector, the finest Trojan warrior, who killed him. Achilles emerged seeking vengeance but, unfortunately, Hector now had his armour, stripped from the dead Patroclus. The divine blacksmith Hephaestus was commissioned to forge some new gear for Achilles, the most spectacular item of which was a shield.

The *Iliad* includes a florid description of the decoration on this shield. It was circular and, around the edge, Hephaestus painted the Ocean, which the Greeks believed to form the boundary of the world.[1] On the face of the shield, identified as the Earth by the ring of the Ocean around it, he depicted various scenes from everyday life in Greece. And in the centre, he placed the Sun, Moon and stars, much like the plan on the sarcophagus of the Egyptian priest Wereshnefer. Homer specifically noted the constellation of the Great Bear (colloquially called the Plough in the United Kingdom and the Big Dipper in the United States). Unlike the other stars that dipped below the horizon for part of the year, he said the Great Bear never bathed in the waters of the Ocean.[2] This is a poetic way of saying that it was visible throughout the year.

The Homeric World Picture

Unlike the landlocked Persians, the maritime Greeks fancied that the sea formed the edge of the world. When the sun set, it fell beneath the waves rather than moving behind a mountain as in the Babylonian story of Gilgamesh.[3] Homer provided a few more details about his world picture elsewhere in the *Iliad*. The sky was a dome on which the stars moved, rather than a home for the gods. As everyone knows, members of the Greek pantheon dwelt on Mount Olympus. Underground, he placed Tartarus, the prison in which the Titans languished, as far below the Earth as the sky was above it.[4]

Later Greeks took his vision very seriously. As we'll see, early Greek maps followed Homer's lead by showing the Earth as circular, with the Ocean running around the edge. Still, even if some geographical kernels can be winnowed from Homer's epics, he was much more concerned about values. Just as the Babylonians, Persians and Egyptians imagined that the constitution of the world followed the contours of their political and religious culture, archaic Greeks wrote their ethics into the structure of the universe. Homer's epics portray a Bronze Age warrior aristocracy bound by rules of honour and shame. The highest good for a Homeric warrior was public esteem, and his greatest fear ridicule or losing face. That's why Achilles refused to fight after he had been insulted by King Agamemnon. He was only tempted back to the battlefield to avenge the death of his friend Patroclus, killed by the Trojan champion Hector.

As Homer told it, the behaviour of heroes like Achilles and Hector adhered to a code of conduct that bound even the gods. It demanded bravery in battle, loyalty to family and respect from one's peers. Since their warrior ethic formed part of the universal moral order, these aristocratic heroes believed that the universe itself validated their violent lives. Earth was merely a stage upon which they performed their daring deeds. By weaving their standards of behaviour into the fabric of reality, the Greek nobility thought they could show their conduct was objectively right.

Neither Homer, the noble elite nor the moral order of the world cared about the lives of ordinary people. The gods' role was to police the rules, but they could not prevent what fate ordained. In a famous scene in the *Iliad*, Zeus contemplates saving one of his sons from Patroclus, even though he is doomed to die, but is warned off such recklessness by his wife, Hera.[5] The Furies, primordial beings who pursued and destroyed those who broke the natural law, were beyond the control of even the gods.

It's no wonder that the religion of Homer's heroes legitimized their lifestyle. He wasn't alone among the Greeks in believing their ethical system inhered within the physical world. For them, the dividing line between the human and the natural, which we take for granted, did not exist. As Cambridge classicist Francis Cornford (1874–1943) put it in the early twentieth century, 'The structure of the world was itself a moral or sacred order because, in certain early phases of social development, the structure and behaviour of the world were held to be continuous with – a mere extension or projection of – the structure and behaviour of human society.'[6] Cornford was saying that, as far as the Greeks were concerned, the way the world worked provided a model for the structure of their society. He could have said the same thing about the Babylonians and Egyptians. Still, while all ancient civilizations agreed that nature encoded a moral order, there was no consensus about what that ought to be. One near contemporary of Homer had no sympathy for the warrior aristocrats and hoped the universe heeded a gentler plan.

The Poet of Helicon

Sometime around 700 BC, a shepherd left his flocks on the slopes of Mount Helicon in central Greece and headed east. When he reached the coast, he took passage to the city of Chalcis on the island of Euboea, just off the mainland. It was the first time he had been at sea. At Chalcis, he joined the funeral ceremony of a local king where, as part of the mourning rituals, he found athletes competing in honour of the deceased. Alongside the sporting events, there was a poetry competition. Later legend pretended that Homer himself had entered the lists. The shepherd from Mount Helicon performed his poem about the origins of the gods. It was the winning entry.[7]

The poet's name was Hesiod, and the poem he sang at the competition is today called *Theogony*, meaning 'the genealogy of the gods'. Together with Homer's epics, the *Odyssey* and the *Iliad*, it is among the earliest surviving examples of Greek verse. *Theogony* is close to being a Greek equivalent to the Babylonians' *Enuma Elish*. It recounts the birth of the Olympian gods, their battles and marriages, and how Zeus ended up as the dominant deity. It begins with the earth goddess Gaia springing spontaneously into existence from the emptiness of the chasm. Shortly afterwards, Uranus, god of the heavens, appears and joins Gaia to produce a brood of new gods. The story then descends into a Freudian nightmare of fathers eating their sons and sons mutilating their fathers.

Years after he won the prize for *Theogony*, Hesiod composed another poem called *Works and Days*. Whereas *Theogony* is about the immortals, *Works and Days* concerns the passing of the seasons and the life of the poet himself. Hesiod was a farmer with a strong sense of thrift and self-worth. He believed that the keys to success were hard work, being hospitable but not charitable, honouring the gods and following the rules. In truth, Hesiod's relative prosperity was due to his father, who was the first of the family to farm on Mount Helicon. But he interpreted his good fortune as down to his own efforts and so entirely deserved. His impregnable egotism led him to incorporate his instincts into a poem that imbued the universe with his personal moral code.

Although they differed markedly in their ethics, there are obvious resemblances between the physical features of Hesiod's and Homer's universes. Even though the two poets do not appear to have directly influenced each other, they both referred to the Ocean as 'back-flowing', by which they probably meant that it ran into itself, that is, it was circular. Stars fell into the Ocean when they disappeared under the horizon.[8] Like Homer, Hesiod placed Tartarus and heaven equidistant below and above the ground.[9] The roots of the Earth reached down towards the underworld, which was surrounded by a wall to contain the Titans trapped within. According to Hesiod, the gods Night and Day shared a house in Tartarus, but they never occupied it at the same time. When Day was out and about, Night stayed home, but she had always left for work by the time Day got back.

The world picture of a disc-shaped earth on solid foundations, surrounded by water and under a heavenly dome, was a commonplace among the Greeks. As well as Homer and Hesiod, other early poets assumed it. The 'Shield of Hercules', once attributed to Hesiod but clearly a later imitation of Homer, describes its titular shield in similar terms to Achilles'.[10] And, in the sixth century BC, Pindar referred to the steel-shod pillars that support the Earth.[11] Yet, that didn't stop a few individuals, moulded by the cultures in which they lived, from developing an unprecedented alternative. The next few chapters will try to explain how that happened.

5

The Origins of Greek Thought: 'Equally distant from all extremes'

'In all history', wrote the philosopher Bertrand Russell (1872–1970), 'nothing is so surprising or so difficult to account for as the sudden rise of civilization in Greece.'[1] Admittedly, Lord Russell was wrong to claim, as he did in the same passage, that the Greeks invented mathematics, science and philosophy. Nor was the appearance of their civilization particularly sudden. But there is a world of difference between the milieu of Homer's epics on the one hand and, on the other, the dialogues of Plato, the logic of Aristotle or the geometry of Euclid (c. 325–c. 265 BC). We have a right to find this remarkable and it does demand an explanation.

In the centuries after Homer and Hesiod, later thinkers like Plato retained the convention of a moral order that undergirded nature and society, but relegated the myths to background colour. They didn't simply assert how things should be but argued for their conclusions. This didn't happen overnight: there were four centuries (a hundred Olympiads, as the Greeks would say) between Homer and Aristotle. It was during this period that the theory of the Globe was discovered and began to displace the traditional flat earth from the Greek world picture.

Law, Politics and Rational Arguments

We can discern a rationale for the appearance of philosophy at the interface between politics and legal practice. The Greeks were a remarkably litigious people, proud of their laws and happy to exploit them.[2] To prevail in a dispute, you had to win your case by pressing for a particular interpretation of the law and applying it to your circumstances. In Athens, cases were heard by a jury of perhaps five hundred citizens, selected by lot, whom plaintiffs needed to convince of the justice of their cause. The sophists, specialists in rhetorical techniques to sway a crowd, taught their students how to construct an argument for or against almost any proposition. In this way, disagreement and debate became institutionalized and, just as importantly, a route to influence and wealth.

The skills necessary to prevail in court translated into the political arena. City states, such as Athens, Thebes and Sparta, each had their own forms of government. Aristotle's students collected 158 constitutions in an effort to establish the best way to run a polity, while Plato's two longest works, the *Republic* and the *Laws*, dealt with the same issue.

Practical politics could determine the fate of entire populations. In his *History of the Peloponnesian War*, his account of the war that raged between Athens and Sparta from 431 BC to 404 BC, Thucydides (c. 460–400 BC) illustrated how demagogues could sway the people with fine speeches and forensic argument. On one notorious occasion, the Athenian assembly voted to put the citizens of a defeated city to the sword. The next day, after further debate, they rescinded the order and sent a fast galley to prevent the massacre.[3] It is a central theme of Thucydides' narrative that the decisions made by the assembly had a direct bearing on the course of the war and the fate of Athens itself.

Notwithstanding their respect for the gods, the Greeks' political systems were secular. The priesthood comprised wealthy citizens rather than a professional cadre. What made Greek politics different from that of empires like Persia and Macedonia was its participatory nature. While some city states were tyrannies, many distributed political authority to a large group of people, usually the landowning gentry, but sometimes, as was the case in Athens, to all male citizens. This dispersal of power derived from the way that the Greeks organized their military. The core of most cities' armed forces was the phalanx, a formation of armoured spearmen called hoplites drilled to keep close formation and fight as a unit. Each hoplite was expected to maintain his equipment and attend regular training. In return, any citizen wealthy enough to afford arms and armour demanded a share in the control of the city.[4] A city state like Thebes could field several thousand hoplites, all of whom wanted a vote in the city's assembly and were eligible for selection to executive positions.

Athens went a step further. Its military capability rested on its navy of galleys: large vessels with three rows of oarsmen on each side. In battle, the crew propelled these triremes into the enemy's ships to hole them below the waterline with bronze rams, or else board them to fight hand-to-hand. The oarsmen of Athenian galleys were not slaves but free citizens. Because they did not need to bring their gear, many more people could afford to row in the navy than fight in the phalanxes. Athens spread political power to all its male nationals, however poor, because they were needed in times of war to man the fleet upon which the city's power depended.[5] That meant debate and argument were more central to political life than they were in other cities. Since politics was open to a wide range of people, those wanting to push particular policies had to persuade the majority of their fellow citizens. A talent for rhetoric was an essential qualification for this. Greek politics 'put a premium on skilled speaking and produced a public who appreciated the exercise of that skill'.[6] There was but a small step between arguing about taxes or war and ruminating about broader questions such as the best way to live.

It sometimes feels as if philosophy today is the preserve of academics in ivory towers considering esoteric conundrums that few of us can understand, let alone feel are relevant to our lives. Greek philosophy wasn't supposed to be like that. The members of the various schools, such as Platonism, Stoicism and Epicureanism, were united by their belief that their doctrines should provide a broad and overarching model for life.[7] A true philosopher's goal was tranquillity and contentment. Followers of Plato sought to achieve this condition through contemplation of the highest good. Another school, the Stoics, worked towards acceptance that the universe was ultimately working as it should, while Epicureans taught that happiness followed from the realization that the philosopher could not truly be harmed by anything that happened to him. Not all Greek thinkers emphasized the supernatural, but most found room for God or gods in their systems of thought. What distinguished the philosophical schools from popular religion and political movements was their elitism. Philosophers admitted that their teaching was not for everyone but only for those with the requisite intellectual aptitude.

Philosophy was most concerned with ethics, broadly construed, so raised issues like 'What is the best way to govern the city?', 'Where does the law come from?' and 'What is the link between politics and virtue?' When laws derived from an absolute monarch or emanated from a holy book, these puzzles didn't matter so much. For the Greeks, where power was dissipated among many individuals, they were essential. The first records of people asking these questions concern a group of thinkers in the sixth century BC, a couple of hundred years after Homer had composed the *Iliad* and the *Odyssey*. They came from the city of Miletus at the southwestern corner of Asia Minor, where the revolt against the Persians would ignite a century later.

The Milesians

By tradition, the founder of Greek philosophy was a man who lived in the early sixth century BC called Thales of Miletus. Unfortunately, we know almost nothing about him. Most of the factoids that can be gleaned from later sources are unverifiable, and several are obvious fables. Plato tells us one famous tale, but it's hard to take it seriously:

> As Thales was studying the stars and looking up, he fell into a well. A Thracian servant girl with a sense of humour, according to the story, made fun of him for being so eager to find what was in the sky that he was not aware of what was in front of him.[8]

Thales' enthusiasm for astronomy features in other stories about him. He is said to have predicted an eclipse of the Sun, which occurred on 28 May 585 BC, although this was certainly beyond anyone's proficiency at the time. At best, he could have been aware of Babylonian lore about when eclipses were more likely to occur, based on correlations derived from centuries of observations.[9] Thales might have been influenced by eastern ideas in other respects as well. Allegedly, he believed water to be the underlying substance from which the world is constituted, perhaps reflecting the Babylonian *Enuma Elish*, in which creation began with the mixture of sweet and bitter waters.[10] However, Homer is a more likely source, since he stated that the god Oceanus was the font of all things.

While we know very little about Thales, there is slightly more information about his successors, such as Anaximander (c. 610–c. 546 BC), also from the city of Miletus. He thought that the ultimate basis of existence was 'the infinite', which might mean the primeval void of Greek creation myths.[11] After all,

Hesiod had begun his genealogy of the gods with an empty chasm, from which the earliest deities spontaneously emerged. The last of the early sages from Miletus, Anaximenes (*c.* 586– *c.* 526 BC), concluded that air was the essence of all things. Realizing that we must breathe to stay alive, early Greeks associated air with the soul.[12] Anaximenes extended this equivalence, perhaps implying that the universe itself was a living thing.

Despite their differences, on one matter the three Milesians were united: they all thought the Earth was flat. However, that was about as far as their agreement went. Thales, believing water was the primary substance, said that the Earth floated on the deep like a log. Aristotle, writing a couple of centuries later, was scathing about this theory. He asked, sarcastically, what all this water is supported by.[13]

Anaximander proposed that we live on the flat surface of a squat cylinder, one-third as high as it is wide. One ancient source compares this to a column drum.[14] This is one of the slices of marble that are piled on top of each other to make the columns that hold up many Greek temples. Anaximander didn't think the Earth needed to be supported by anything. He said it was established at the centre of the universe, equally distant from all extremes, and held in place by equilibrium. Aristotle was rude about this idea as well. He compared it to a hungry and thirsty man, equally far from food and a drink, who couldn't make up his mind which to reach for first.[15]

The third Milesian, Anaximenes, said that the Earth rested on air. Because it was flat, it could float on its cushion of wind, rather like the way that a breeze can keep an umbrella aloft. He may also have postulated that the Sun hides behind a great mountain to the north at night: an idea we first heard about in the *Epic of Gilgamesh*.[16]

The Milesians' status as the founders of the western intellectual tradition is unassailable, even if their thinking was strongly

influenced by Homer and the Babylonians. Some modern scholars, such as the twentieth-century philosopher Sir Karl Popper
(1902–1994), have claimed that they represent the earliest stage
of the scientific mentality. Popper called their proliferation
of new theories 'bold and fascinating ideas'.[17] He was more
impressed than Aristotle by Anaximander's suggestion that
the Earth stands in balance at the centre of the universe, exu-
berantly calling it 'one of the boldest, most revolutionary, and
most portentous ideas in the whole history of human thought'.[18]

For Popper, the Milesians' defining feature was rationality
and a critical attitude. Never mind that their ideas were fantastical; the mere fact that they tried to understand the world without
recourse to gods was a step in the right direction. However,
their irreligious veneer was thin. While the Milesians stopped
using the names of the traditional members of the Olympian
pantheon, they didn't downgrade the underlying moral order of
the universe. Indeed, the Greek word commonly used by philosophers to describe the organizing principle of the world was
kosmos, which literally means 'order'. Originally, this referred to
an ethical framework or government. Later, it acquired a more
concrete definition, which roughly corresponds to the 'universe',
implying reality is an ordered whole that is subject to rules that
we can understand.[19] The elucidation of the laws that govern the
universe, whether carried out by modern physicists or ancient
philosophers, is called cosmology or 'the study of the cosmos'.

Early Greek Astronomy

The stories that Thales predicted an eclipse and fell into a well
while stargazing illustrate how closely the Greeks linked philosophy with astronomy.[20] However, contrary to the rumours
of Thales' astronomical prowess, knowledge about the heavens
among the Greeks when he was alive was primitive, certainly

compared to the Babylonians. Homer hadn't even been aware that the morning and evening star are the same entity, the planet Venus, something the Sumerians figured out before 2000 BC.[21] By around 1000 BC, the Babylonians had identified the planets visible to the naked eye (that is, Mercury, Venus, Mars, Jupiter and Saturn). The earliest surviving Greek text to name all of them dates from about 350 BC, although they had known there were five at least fifty years previously.[22] Mercury, in particular, is tricky to spot as it sticks close to the Sun and isn't especially bright. That the Greeks inherited astronomical wisdom from the Babylonians is beyond doubt: the Greek names of some planets and constellations show clear affinities to their eastern antecedents. For example, they sometimes called Saturn the 'star of the sun', an appellation attested in cuneiform sources. And the Graeco-Roman goddess Aphrodite/Venus is often associated with the Mesopotamian Inanna/Ishtar, all of whom are identified with the same planet. The Greeks themselves credited the Babylonians with celestial observations and the basic instruments to carry them out.[23]

There were plenty of avenues by which eastern science could have penetrated Greece. It's possible the process was helped along by the Persian occupation of Ionia, including the city of Miletus itself, in the late sixth century BC. This meant that Mesopotamia and the Ionian cities of Asia Minor were all part of the same empire. Western mercenaries served in Persia's armies, and merchants from all over the Mediterranean were a familiar sight in the ports of the Levant.

In common with Homer, early Greek astronomers assumed that the sky was a dome over the Earth. The stars attached to its inner surface appear to orbit each day anticlockwise around an axis that leans north. Nowadays, we can use Polaris to identify due north since it sits almost exactly at the centre of the heavens' rotation.[24] As Homer knew, some stars, located in the northern part of the sky, are always visible as they circle the Earth's axis. Most others rise and set as their motion takes them above and below the horizon. Some are only visible for that part of the year when they are above the horizon at night. The Sun, Moon and other planets participate in the daily rotation of the celestial dome. But, unlike most of the stars, which stay in the familiar patterns of the constellations, these wandering lights (*planetes* is Greek for 'wanderers') also move slowly clockwise across the sky in a narrow band against the background of the fixed stars.

Greek astronomers assimilated what they could from Babylonian sources. However, this didn't teach them how the cosmos was put together or the way it operated. That was a question for philosophers, but it would take decades before they were able to provide coherent explanations for any astronomical occurrences. The work was done by members of a group of thinkers that today we call 'the Presocratics', for no better reason than that they lived before and during the career of the celebrated Athenian Socrates (470–399 BC). Besides the Milesians, they include a dozen or so other thinkers who were active in the sixth and fifth centuries BC.

The Presocratics fascinate modern scholars. The amount of research devoted to them dwarfs that on later Greek philosophers, excepting only the trinity of Socrates, Plato and Aristotle. None of their writings survive complete, so we have to make do with quotations embedded into later works. It's the combination of their antiquity and fragmentary remnants that makes them an irresistible challenge to academics.

Unless you are a specialist, it is hard to keep track of the cavalcade of Presocratics with their unpronounceable names and strange doctrines. Luckily, only two of them have significant parts to play in our story: Parmenides of Elea (*c.* 515–450 BC) and Anaxagoras of Clazomenae (*c.* 500–428 BC). In common with all their contemporaries, they thought the Earth was a disc. However, they also developed theories about the Sun and Moon that would lead later thinkers to reconsider the shape of the Earth. Socrates probably met Parmenides and Anaxagoras, and we'll get to know all three in the next chapter.

6

The Presocratics and Socrates: 'Floating on air'

In the early fifth century BC, the most famous philosopher in the Greek-speaking world was Parmenides of Elea. He came from southern Italy but appears to have travelled widely in the course of his life. Parmenides was a truly revolutionary thinker. He taught that existence never changes and everything we perceive is an illusion: a doctrine he expounded through the medium of poetry. Here was a philosopher willing to accept the truth of conclusions he had reached through rational deduction, even if they conflicted with everyday experience. He had concluded that change was impossible and he would not allow his senses to overrule his reasoning.

Parmenides also described the visible world of everyday experience. This seems like an odd thing for him to do, given he asserted it was a simulacrum that deceives us, and we don't possess enough of his poem to know why he did. Still, in this context, he's said to have been the first person to say the Earth is a sphere.[1] Unfortunately, although fragments of Parmenides' poem survive, there's no mention in them of the shape of the Earth. All we have is the single third-hand reference by Diogenes Laertius, a biographer of Greek philosophers writing in about AD 200. Diogenes attributed the first mention of the Globe to other thinkers too and didn't try to reconcile the inconsistencies in his sources. As for Parmenides, it is likely that someone misinterpreted his words.[2] In an extant scrap of his poem, he states that the cosmos as a whole is spherical, this being the perfect shape.[3] Since a sphere is symmetrical in all directions, it never alters however you look at it. This is consistent with Parmenides' principle that change is impossible. It seems that a misunderstanding over whether he was talking about 'the world', as in the whole universe, or just the Earth itself, accounts for the testimony that he said the Earth is a sphere. Another possible source of confusion is that the Greek word for 'round' could mean both 'circular' and 'spherical'.[4] We'll encounter these ambiguities again, since they exist not only in Greek, but Latin and English, too.[5]

The Light of the Moon

Parmenides placed the disc of the Earth at the centre of the spherical universe, which meant the heavens surround us on all sides. One implication of this idea is that, even after the sun has set, it continues to shine from below and can thus illuminate the face of the Moon. As Parmenides put it, poetically, the Moon is 'a light by night, wandering around the Earth with borrowed light, ever gazing toward the rays of the sun.'[6] So, while he didn't say the Earth is a sphere, he probably was the first person to correctly state that the Moon reflects light from the Sun.[7]

Admittedly, in another passage, he said that the Moon is composed of fire and earth, which suggests that it also glows with a fiery light of its own.[8] This might be a reference to the dim red colour the Moon can exhibit during a lunar eclipse. Puzzles like this are all too common when studying the Presocratics. Even when you can determine exactly what they said (which you usually can't), it is another matter to figure out what they meant.

According to Plato, Parmenides visited Athens late in life.[9] If this is so, he would have undoubtedly dropped in to see Anaxagoras of Clazomenae, the city's most prominent philosopher at the time. From the scanty and late evidence, it seems likely that Anaxagoras was born to a well-to-do family on the Ionian coast of Asia Minor. He moved to Athens as a young man and quickly gained a reputation as an avant-garde philosopher by writing a book that postulated that the Sun was a lump of molten metal rather than a god. He downgraded it to a ball of fire the size of the Peloponnesian peninsula, which is about 160 kilometres (100 mi.) across.[10] This was extremely controversial. The Greeks thought the heavenly bodies were divine, so reducing them to the status of hot stones was potentially blasphemous. However, when a large meteorite fell out of the sky in northern Greece in 467 BC, it seemed to be corroboration for Anaxagoras' theory and made him into something of a celebrity. In any case, while he had a reputation as a sacrilegious thinker and was sceptical about the traditional Olympian gods, Anaxagoras was no atheist. He believed the cosmos was governed by a divine mind, which performed the function of the moral order assumed by the poems of Homer and Hesiod.

In Athens, Anaxagoras joined the circle of Pericles (c. 495–429 BC), the populist ruler of the city. At some point, Pericles' political opponents tried to damage him by promulgating a decree against Anaxagoras, using his contentious ideas as pretext. He had to flee the city into exile. It's said that, on being asked if he missed the Athenians, he responded that he didn't, but they missed him. In any case, he wasn't concerned about dying in exile since the road to hell was much the same wherever you start.[11]

Even if they never met, Anaxagoras would have known of Parmenides' poem. Both philosophers adopted the spherical universe, with a flat earth at its centre, and presumed that the Sun, Moon and planets move around and underneath it. Parmenides had already noted one consequence of this: that the Moon reflects light from the Sun. Anaxagoras agreed, and the doctrine became associated with him in the minds of later writers.[12]

This model of the cosmos led Anaxagoras to develop a spectacularly successful explanation for eclipses. Armed with the insight that the Sun can shine on the Moon from below the horizon, he asked what might happen when Earth got in the way. Furthermore, since the Moon could be above the Earth during daylight hours, albeit usually rendered invisible by the glare, it could get in the way of the Sun. He correctly deduced that a solar eclipse is caused by the Moon passing between the Sun and the Earth, while a lunar eclipse occurs when Earth obscures the Moon's disc from the Sun.[13] This meant that the Earth's shadow was visible when it passed across the face of the Moon.

The most shocking implication of Anaxagoras' theory was that eclipses were not supernatural portents. In the summer of 431 BC, shortly after the start of the Peloponnesian War between Athens and Sparta, there was an almost total eclipse of the

Diagram of an eclipse of the Sun, with the Moon casting a shadow cone, illumination from *De sphaera* (1260).

Sun, such that the stars became visible by day. The Athenians were terrified, but Pericles, recalling discussions with his friend Anaxagoras, explained to the frightened men that it was a natural and predictable event that was always going to occur at that moment.[14] Another account of the event said that Pericles used his cloak to demonstrate how the Moon got in the way of the Sun.[15]

Anaxagoras didn't develop his model of the heavens in isolation from the rest of his philosophy. He believed that the divine mind governed the cosmos according to the dictates of reason. For him, finding a natural cause for eclipses was a vote of confidence in divinity, not a way to marginalize it. His thought was an early example of how monotheism has encouraged scientific explanations of the world.[16] Instead of fearing any disturbing event, like an earthquake or occultation of the Sun, as a direct intervention by the gods, he expected the rational principle behind the universe to arrange things so that these events could be understood in terms of physical processes.

Without detracting from the achievement of Anaxagoras, we should note other aspects of his cosmology have not stood the test of time. He believed there were additional dark objects in the sky that could cause eclipses when they passed in front of the Moon or Sun. He took the meteorite that landed in Greece to be one of these objects. And, like the Milesian Anaximenes, he imagined that the Earth rested on a cushion of air. When this air seeped into the hollows within the ground, it could cause tremors and quakes.[17]

The awareness that the Moon reflects light from the Sun allowed astronomers to deduce that it had to be spherical. This is the only shape consistent with the Moon's appearance as a crescent when it waxes and wanes. Meanwhile, the dark penumbra during a lunar eclipse gives a glimpse of what the edge of the Earth looks like from space. It is clearly rounded. That said,

Anaxagoras thought that Earth was a flat disc, whose rounded edge we can see during a lunar eclipse. It was perfectly reasonable for him to hold this view. Since he had no answer as to why we would not fall off if the Earth were a sphere, it's hard to see how he could have taken any other position.

One of Anaxagoras' followers was an otherwise obscure figure called Archelaus (*fl. c.* 450 BC).[18] What little is known about him shows that he built on his master's ideas. In particular, he modified the Earth's shape from a flat disc to a round bowl. His reason for making this change was that 'the Sun does not rise and set at the same time for all men as would inevitably happen if the Earth was flat.'[19] Evidently, the seafaring Greeks had realized that the hours of daylight were not the same in every part of the world and that this observation could not be explained by a flat earth.

At the base of the crucible formed by the Earth, Archelaus placed a marsh that he said was the source of life. Water flowed down into the bog where heat separated it into warm air, which rose to the skies, and cool earth from which animals emerged. Thus, the cold elements of earth and water tended to settle in the centre, while the hotter elements of air and fire naturally moved upwards.[20] The fragmentary evidence for Archelaus' thoughts makes it difficult to follow how he imagined these processes took place, but we'll see later how they might have inspired Aristotle while he ruminated over his own world picture.

Socrates in the Clouds

According to a suggestive tradition, Archelaus was the tutor of Socrates. We should take this with a pinch of salt because Greek biographers liked to assert that famous philosophers learnt their vocation at the feet of other famous philosophers. But unless Socrates was an autodidact, he would have had a teacher and the claim that this was a nonentity like Archelaus is inherently more plausible than trying to link him to a major figure.

Socrates is far more famous than any of the thinkers who preceded him. The god Apollo, speaking through the oracle at Delphi, declared he was the wisest man in Athens. Most of his fellow citizens just found him irritating. He was a man of modest means, a stonemason by trade, but with good social connections. He had a wife, whom the ancient sources show a characteristic lack of interest in, and at least three children. Many surviving marble busts portray him as ugly, with a pug nose, making him instantly recognizable.

The esteem in which we hold Greek philosophers wasn't necessarily shared by their contemporaries. A typical Athenian thought they were suspect chancers, with their newfangled ideas and weird ratiocinations. Socrates was as bad as any. By 423 BC, he was notorious enough to feature in a comedy by the playwright Aristophanes (c. 446–c. 386 BC), performed during the festival of Dionysius.

The play was called *The Clouds* and its script survives in full. It features a fictional Socrates as an eccentric teacher driven to distraction by an obtuse student. More dangerously, he is shown rejecting the agency and even existence of the gods. For example, he denies that thunder is caused by Zeus and says it is the result of the forced evacuation of air from clouds, giving the cue for a series of fart jokes.[21] Elsewhere in the play, Aristophanes threw in loads more sacrilegious remarks and put them all into the mouth of poor old Socrates. He still remembered how the play misrepresented him when, a quarter of a century later, he was on trial for his life.

Incidentally, *The Clouds* contains some clues about what people thought about the shape of the world in the fifth century BC. Aristophanes made two references to the issue. One of the characters mocks philosophers for thinking that the sky

Bust of Socrates,
1st century AD, marble.

is a solid dome like a bread oven.[22] Later in the play, Socrates echoes Anaxagoras' view that the Earth is held up by air, calling on the sky to 'hear my prayer, O Lord, O King, O boundless air on whom the Earth supported floats.'[23] The script ridicules these ideas. If people at the time had been discussing something as crazy as the Earth being a sphere, this would have been a target for satire that Aristophanes would hardly have been able to resist. However, there is no mention of it in his plays or any other fifth-century source. Incidentally, *The Clouds* bombed with the critics, coming third in a field of three in the competition at that year's comedy festival.

The Death of Socrates

The second half of the fifth century was dominated by the Peloponnesian War, an existential struggle between Athens and Sparta. While the catalyst for the conflict was a dispute over the city of Potidaea in 432 BC, the root cause was that fighting each other was just something Greek city states did. Homer's warrior ethic was alive and well among the upper classes, who saw war as a road to glory and wealth. During the first part of the conflict, Socrates fought as a hoplite, a role that he performed with notable gallantry.

The philosopher was too old to fight long before the struggle ended, but he served in Athens's democratic government in 406 BC. He outraged his colleagues by declining to vote for an unconstitutional trial of military leaders who had abandoned the bodies of the dead after a battle. The trial went ahead anyway and the defendants were put to death. Then, in 404 BC, after Athens had been finally defeated, an oligarchy replaced the democracy. The junta ordered Socrates to detain a man he believed was innocent. He refused. In a few short years, he had offended both the democratic party and the oligarchs. In 399 BC, his enemies arranged for him to be charged with corrupting the young and impiety, which carried the death penalty.[24]

It is plain that the accusations against him were trumped up. Socrates had managed to alienate every faction in Athens, and his prosecutors wanted to make an example of someone who had no one to protect him. The accusation that he was 'corrupting the young' came about from allegations that he taught his students to deny the traditional gods. At his trial, Socrates didn't help himself, going out of his way to antagonize the jury and making little effort to defend himself against the charges. He blamed his bad reputation on his portrayal in *The Clouds* and complained that his accusers had confused his respect for the gods with the atheism of Anaxagoras. Although Anaxagoras had been dead for decades by 399 BC, his book was still available cheaply at bookstalls in the marketplace of Athens. Socrates was inevitably found guilty and sentenced to death. His friends urged him to escape the jail where he was awaiting execution, but he refused to abandon Athens, even though the city had wronged him. He died by drinking a cup of hemlock given to him by the executioner.

These events form the dramatic setting of a series of dialogues by Plato. The *Apology* recollects the defence speech given by Socrates at his trial, while, in the *Crito*, Socrates admonishes his friends, who are begging him to save himself by going into exile. Finally, the *Phaedo* takes place in the prison where Socrates spent his final hours.

For the purposes of our investigation into the origins of the Globe, the *Phaedo* is a watershed: it's the earliest extant record of anyone saying that Earth is a sphere. Socrates raises this question in the dialogue but, in real life, it is something that he almost certainly never considered. Rather, he is a vehicle for Plato's own opinions a generation later. Before we consider the contents of the *Phaedo* in more detail, we need to meet Plato and find out how he might have heard that the Earth is round.

7

Plato: 'Flat or round, whichever is better'

During a lecture given in the Scottish city of Aberdeen in the late 1920s, Alfred North Whitehead (1861–1947) characterized European philosophy as a 'series of footnotes to Plato'.[1] He meant that Plato introduced many of the fundamental concepts in Western thought, setting the agenda for later thinkers. The Globe is a typical example. It is first recorded in one of Plato's dialogues, but he left it to his successors to explore the implications of this unconventional conjecture.

The Life of Plato

During his career, Plato published over thirty dialogues, all of which survive to this day. As well as being philosophically profound, they are paragons of Greek prose style. In all except the last, Socrates is one of the interlocutors in the conversation, talking to various real people who would have been alive at the time the dialogue is set. The Presocratic, Parmenides, stars in one, while Aristophanes, the comic playwright Socrates blamed for mischaracterizing him, features in another. Despite their verisimilitude, the dialogues are fictional. Modern scholars continue to debate the extent to which the ideas they contain are Plato's own or represent what he had been taught by his master. Nonetheless, there is a consensus that the early dialogues reflect the teaching of the historical Socrates, whereas this is not the case for those written later in Plato's life, even though Socrates remains a character in them.

Plato came from an upper-class family who were members of the cabal that ruled Athens after its defeat by Sparta in the Peloponnesian War. Though he was too young to join his relatives in government, an aristocrat like Plato was expected to participate in the political life of Athens once he came of age. However, before he was thirty, he had experienced the rule of both a vicious oligarchy and a populist democracy. The oligarchs appalled him, but it was the democrats he held responsible for the trial and execution of his master, Socrates, whom he admired as the best of men. Eschewing politics, he refused to join any faction and devoted himself to philosophy. He founded a school on the outskirts of Athens next to a sacred olive grove associated with Academus, an obscure ancient hero. Posterity named his foundation the Academy.

Plato was unable to avoid all political involvement. In 387 BC, when he was already a famous teacher, a friend persuaded him to travel to Sicily to act as an adviser to Dionysius (*c.* 432–367 BC), the tyrant of Syracuse. Syracuse was a thriving Greek colony that had been the scene of a catastrophic Athenian defeat during the Peloponnesian War about three decades previously. Dionysius turned out to be a boorish thug with no interest in philosophy, but Plato did make some friends while he was visiting the western Mediterranean. Among them was a devotee of Pythagoras called Archytas (*c.* 410–*c.* 350 BC), who was a renowned general in the city of Tarentum at the heel of Italy.[2] Meeting Archytas might have piqued Plato's interest in Pythagorean thought and, in any case, the two men remained friends and corresponded for years afterwards. The dialogues that Plato wrote later in his life seethe with Pythagorean themes, although he explicitly mentioned Pythagoras himself only once.[3] Given that the earliest surviving reference to the Earth being a sphere is in a dialogue that Plato wrote after returning from his sojourn in Italy, it is worth investigating whether this was one of the nostrums he picked up from the Pythagoreans he met there.

Pythagorean Dreams

Very little is known for certain about Pythagoras. He was born around 570 BC on the island of Samos in the Aegean Sea and moved to the Greek colony of Croton in southern Italy as an adult. While there, he preached an austere way of life and collected a group of disciples. His followers became politically powerful in Croton and nearby cities but were ultimately expelled by the citizens, who resented their influence. After their master's death, many Pythagoreans drifted back to Greece where they preached their creed in Thebes and Athens. By the mid-fifth century, they had become a well-known sect of thinkers. Nothing written by Pythagoras survives, and modern scholars doubt he wrote anything at all. Later Pythagoreans back-projected their own opinions onto him, making it nearly impossible to thresh later accretions from the kernel of his original creed.[4]

Pythagoras is best known for the eponymous theorem connecting the dimensions of a right-angled triangle. However, Babylonians and Egyptians had been aware of this centuries before he lived. It is likely he taught that numbers have metaphysical properties and are, at some level, the building blocks of reality. Of the other tenets attributed to him, scholars have concluded that he probably believed in reincarnation. It was claimed that Pythagoras could recognize people even after their souls had transferred to another body, and that he once intervened to prevent a dog from being beaten because the cur had been a friend in a past life. Aristotle said he forbade the eating of beans, although quite why he'd do this is a mystery.[5]

Numerology proved a more productive doctrine than reincarnation, leading later Pythagoreans to make significant mathematical advances. They developed a musical scale and deduced the relationship between the length of a vibrating string and the tone it produces. Characteristically, they credited these discoveries to Pythagoras himself. He's said to have deduced musical intervals as he wandered past a blacksmith, by listening to the notes produced by hammers of different sizes clashing with the anvil.

Despite their prowess as mathematicians, the Pythagoreans were mystics rather than rationalists. They aimed to attune themselves to the universe by understanding its vital resonances. One of their signature doctrines held that the movement of the stars produced a harmony called 'the music of the spheres'. Aristotle poured scorn on the idea, but it remained influential until the seventeenth century.[6]

Our old friend Diogenes Laertius claimed that Pythagoras was the first to say the Earth is a ball.[7] Needless to say, there is no good evidence for that. Recall that Diogenes also wrote that Parmenides had been the earliest to teach this. He claimed the honour for the poet Hesiod as well, not to mention Anaximander of Miletus.[8] Diogenes didn't appear to have been troubled by contradictions in his sources, even though he must have known they can't all be right. In fact, when it comes to the Globe, none of them were. Parmenides said the universe is an orb, but not the Earth. As for Hesiod, like everyone else in the eighth century BC, he instinctively thought it was a disc. We should treat the two occasions that Diogenes said Pythagoras knew the Earth is round with the same scepticism.[9] It is more likely that a certain Philolaus (*c.* 470–*c.* 385 BC), one of his later followers, sowed the seed that led to the theory of the Globe.

Earth among the Stars

Philolaus was a contemporary of Socrates, born in southern Italy, perhaps in Croton, where Pythagoras had spent much of his life. As the political environment in Italy became more hostile, Philolaus moved to Thebes in Greece but appears to have returned to Italy before he died. A late tradition makes him the teacher of Plato's Italian friend Archytas.

Philolaus wrote at least one book, called *On Nature*, from which later authors quoted several excerpts. For a long time, scholars assumed these were inventions, but careful work by the German philologist Walter Burkert (1931–2015), in the 1960s, has led to a consensus that about eleven of the fragments are genuine. Burkert concluded that the other quotations attributed to Philolaus are spurious and date from long after his lifetime. In short, the evidence we have about him is fragmentary, at second hand and possibly corrupt.[10] This is typical for research on Pythagorean thought, and classicists wouldn't have it any other way.

The case that Philolaus put forward the idea of a spherical Earth is circumstantial but coherent, so it is worth trying to assemble the evidence, starting with one of the authentic fragments of his book. This reads, 'The first thing fitted together, the one in the centre of the sphere, is called the hearth.'[11] In this statement, Philolaus was saying that the universe is a ball with some sort of fireplace in the middle of it.

Another major source for the cosmology of Philolaus is a book by Aristotle called *On the Heavens*, which dates from about 345 BC. We'll be looking more closely at this masterpiece of ancient science in the next chapter. In the book, Aristotle criticized the theories of a group of Italian Pythagoreans.[12] He didn't name Philolaus, but it's safe to assume that is whom he was talking about because the system he described is consistent with the genuine fragments of Philolaus' book.

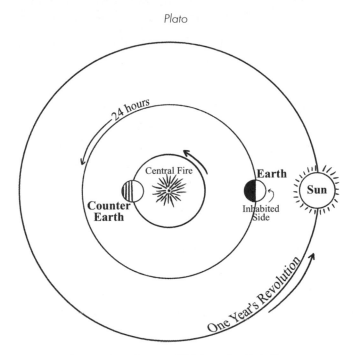

The cosmological system of Philolaus, after M. A. Orr,
Dante and the Early Astronomers (1913).

According to the world picture of Philolaus, the constellations are sprinkled over the outermost shell of the universe. At its core, there is a 'central fire' or 'hearth', orbited by the Sun, the Moon and five other planets. Earth also revolves around the central fire, completing its orbit every 24 hours. By contrast, most ancient Greeks, including Aristotle, assumed that the Earth was stationary and the entire universe rotated around it daily. Philolaus also postulated a further mysterious body, the counter-earth, which sits directly opposite our Earth on the other side of the central fire. Both the counter-earth and the central fire are invisible to us because the inhabited side of the Earth always faces away from them.[13]

It is tempting to imagine that, with this model, Philolaus anticipated the modern picture of the solar system. After all, he said Earth circles the hearth like one of the stars. But he did not say that Earth orbits the Sun. On the contrary, he placed the Sun further out, so that, for part of the time, the Earth's orbit means it faces away from it. In other words, Philolaus was explaining night and day, making him the earliest example of someone hypothesizing the diurnal rotation of the Earth.[14]

Philolaus' explanation of the daily cycle of light and darkness is elegant. However, there's no empirical evidence for the existence of other aspects of his cosmos, like the central fire or the counter-earth. Aristotle was scathing about this kind of reasoning, accusing the Pythagoreans of inventing theories to fit their preconceptions. He thought the counter-earth might have been intended to bring the total number of heavenly bodies to ten, a sacred number to the Pythagoreans.[15] However, this could just be a bit of snark on Aristotle's part. The Pythagoreans had plenty of sacred numbers, so they didn't need exactly ten heavenly bodies. Seven planets would have done just as well. It's more likely that Philolaus invented the counter-earth to keep the universe in balance, so that the Earth's off-centre position didn't throw reality out of kilter. It is best to see Earth and counter-earth as forming a rotating dumbbell, perfectly poised with the central fire between them.

None of this proves Philolaus believed the Earth is a sphere. Neither the extant fragments of his work nor the paraphrase in *On the Heavens* explicitly say so. Aristotle cited his world picture as an example of someone claiming that Earth moves rather than it being round. However, he did mention in passing that the Earth and counter-earth are only a hemisphere away from the central fire. That implies that they are themselves spherical. He also noted that the Pythagoreans treated Earth as one of the stars. Since he had already said that all the stars, including the Sun and Moon, are spheres, that's a hint that they said the Earth was one too.[16] Besides, a spherical earth seems consistent with the cosmology of Philolaus. It is hard to imagine a flat earth orbiting the hearth in the way he described as it would spend half of its time upside down.

Overall, despite the frustration that we nowhere find an explicit statement that Philolaus said the Earth is a sphere, he is our best candidate for the first person to suggest it.[17] If he did, it was an impressive intuition, even if more akin to a lucky guess than a reasoned hypothesis. Perhaps he ascribed the spherical Earth to Pythagoras himself, imputing his world picture to his revered predecessor as recorded by Diogenes Laertius centuries later. It would also explain where or whom Plato heard the idea from. If Pythagoreans in southern Italy were discussing the shape of the Earth when he visited around 387 BC, he could have picked it up from them. Diogenes even reported that Plato met Philolaus while he was there.[18] In any case, the earliest extant references to the theory of the Globe turn up in the *Phaedo*, which Plato wrote within a few years of returning home to Athens. In fact, the dialogue alludes to a spherical Earth twice.

Before an Execution

The *Phaedo* is set in the prison where Socrates is waiting to be executed. A few close friends have come to keep him company during his final hours. Phaedo, from whom the dialogue gets its name, is one of the group. Plato himself is ill that day, so cannot attend to his old master.[19] Because he explicitly stated this, Plato wanted his readers to know the dialogue is not an eyewitness account. It was a product of Plato's imagination and he used the scenario as a frame for his own views. This means that the *Phaedo* doesn't tell us very much about what Socrates was really saying at the time of his death. Nor does it reflect the intellectual world of 399 BC, the year Socrates died. Instead, the philosophical themes in the dialogue are Plato's thoughts in the 370s BC, when it was probably written. That said, Plato did care about the authenticity of his dialogues, populating them with real people and locating them in familiar settings around Athens. So, incidental details are more likely to be accurate than the words of the speakers.

The philosophical core of the *Phaedo* is Plato's doctrine on the immortality of the soul. He milked the dramatic potential of a conversation about life beyond the grave as Socrates prepares for execution, while also trying to provide a rational demonstration of his views. Even the part of the dialogue that deals with the shape of the Earth does so in the context of where souls go after death. Plato believed that we are judged for our deeds in life and sent for reward or punishment accordingly. However, except for extreme cases, our souls eventually settle in new bodies, a doctrine that Plato tried to prove by showing we retain some knowledge from our past lives. The topic of reincarnation shows the influence of the Pythagoreans, not least because Plato only began to focus on the problem of the soul after he had returned from Italy. Philolaus even gets a name check in the *Phaedo*, although we learn nothing about his opinions on the subject at hand.

Early in the dialogue, Socrates talks about his own philosophical development: as a young man, he was fascinated by the workings of the universe. However, all the weird and wonderful ideas he learnt about physics had left him more confused than when he started. The bewildering variety of speculation in early Greek cosmology and medicine makes this a fair point. It was also the experience of the historical Socrates. Late in his life, he was scornful of natural philosophers, not least because they concocted a multiplicity of explanations for phenomena and never agreed on anything. From this, he concluded that none of them knew what they were talking about.[20]

As the conversation in the *Phaedo* continues, Socrates mentions he had chanced upon a book by Anaxagoras of

Clazomenae, perhaps the one available at Athenian bookstalls at the time of his trial. In the book, Anaxagoras had explained that there was a divine intelligence behind the world, which had created it to be the best it possibly could be. Socrates was immediately attracted to this idea. He believed that the cosmos was the product of a benign and skilled craftsman. Reading on in the book by Anaxagoras, Socrates wished it would illustrate the perfection of the world by showing how its organization really was for the best. 'I assumed', Socrates says, 'that [Anaxagoras] would begin by informing us whether the Earth is flat or round, and then he'd explain why it had to be that way because that was what was better.'²¹ Alas, his hopes quickly dissipated. He found that even though Anaxagoras thought the world was the product of a mind, he expounded its constitution in terms of natural causes. In contrast, Socrates thought that the order of creation was an ethical rather than a physical question.

For us, the importance of this passage is it contains the very first extant reference to the possibility that the Earth might be a sphere. The Greek word that Plato used can be translated as 'round' as well as 'spherical', but in the context of comparing it to a flat earth, it must refer to a ball. Thus, we know that when Plato wrote the *Phaedo*, the idea of the Globe existed.

Can we also conclude from Plato's text that Anaxagoras mentioned the concept in his book, which was written a hundred years earlier? Some modern scholars believe so. They suggest that Parmenides was the first to propose the Globe (as Diogenes Laertius said) and that Anaxagoras, who thought the Earth was flat, tried to refute him.²² If Parmenides had really said the Earth is round and Anaxagoras had argued against, it would mean that the shape of the Earth was being seriously discussed by 470 BC. However, the evidence is slim. In the last chapter, I explained why I don't think Parmenides said the Earth is a sphere. And, we don't hear of anyone else talking about it for another century. So, it is much more likely that the *Phaedo* reflects questions that were being discussed when it was written, in about 375 BC, rather than at the time of Socrates' death in 399 BC, let alone when Anaxagoras wrote his book decades earlier.

The Globe in Service to Philosophy

Plato was no scientist. He treated cosmology as a branch of ethics or metaphysics. When he first raised the theory of the Globe in the *Phaedo*, he didn't ask, 'Is it correct?' but, 'Is it for the best?' And he ignored any evidence from astronomy that might have supported the hypothesis. This attitude comes through again in the *Timaeus*, a dialogue on the nature of the world that Plato composed towards the end of his life. Intriguingly, there is an ancient tradition that he based some of this dialogue on a book by Philolaus.[23]

In the *Timaeus*, Plato described the universe as shaped like a sphere with the Earth at its centre. It was designed from heavenly templates by a divine craftsman, who moulded chaos into a semblance of order. The dialogue doesn't explicitly mention the shape of the Earth, although it implies that it is spherical. In any case, Plato didn't treat the *Timaeus* as factual and admitted it might not be true – only that it was as likely a story as any.[24]

We've seen that earlier Greeks, like Homer and the Presocratics, imagined that a moral order was woven into the structure of the world. Plato's scheme was more sophisticated, but his aim was the same. He sought to use the organization of the cosmos to illustrate that his ethical teaching was true and not just opinion. After all, if morality is a matter of objective fact, it must have its origin outside our heads. Later readers lost sight of the fictional character of dialogues like the *Timaeus* because Plato's reputation meant that everything he wrote was imbued with near-inerrant authority. His admirers interpreted his world picture as the way things had to be and read the *Timaeus* as a scientific treatise describing a universe permeated with purpose.[25]

We need to bear in mind how Plato used cosmology to score ethical points if we are to make sense of the second mention of the Globe in the *Phaedo*. 'This is what I believe then,' says Socrates, a few pages after the discussion about Anaxagoras. 'In the first place the Earth is spherical and in the centre of the heavens. It needs neither air nor any other such force to keep it from falling. The uniformity of the heavens and the equilibrium of the Earth itself are sufficient to support it.'[26] This remarkable passage is the earliest record we have of someone saying they think the Earth is round. In the dialogue, Socrates speaks these words, but they undoubtedly represent Plato's view.

It is significant that, in the *Phaedo*, Plato placed the Earth at the heart of the universe. He explicitly rejected Anaxagoras' theory that it is supported by air in favour of the idea developed by the Milesian thinker Anaximander, who said the Earth needs no support if it is held at an equal distance from all things. That said, Plato wasn't really interested in the physical implications of Anaximander's proposal. He placed the Earth at the centre of the universe and made it round because he thought this was the 'best' way to arrange things.

As Socrates continues to describe the world in the *Phaedo*, his vision quickly becomes a phantasmagoria. For instance, he makes the shocking admission that we don't live on the surface of the Earth, but in deep fissures far below it. From a distance, the Earth looks like a football (soccer ball) made up of twelve pentagons sewn together, with the entire region known to the Greeks, from the Pillars of Hercules to the far side of the Black Sea, contained within an indentation encompassed by one of them.[27] When we look at the sky, we aren't seeing the true splendour of the Moon and Sun, which can only be experienced from the surface of the Earth far above. After death, the souls of the blessed ascend to dwell up there, where everything is perfect and of beautiful colours. Mortals abide in this terrestrial paradise too, some on islands in the air, where they can directly perceive the divine nature of the heavenly bodies. These inhabitants are free from disease and enjoy very long lives. The purest souls soar even higher to the indescribable bliss of heaven. Plato placed Tartarus at the Earth's core: a pit into which the most wicked of the dead suffer for eternity.

None of this is 'scientific'. The world picture is an ethical hierarchy from low to high. Tartarus sits at the bottom while humans inhabit an intermediate level, still deep underground, deluding themselves into thinking they are higher up the moral ladder. Above them, on the surface of the Earth, the gods and blessed reside together. Not even Plato expected his readers to take all this literally. Socrates winds up his story with the words, 'No reasonable man ought to insist that the facts are exactly as I have described them.'[28] The story in the *Phaedo* is just one of the allegories that Plato used in his dialogues to elucidate his belief that our everyday world is a dim reflection of a greater reality that most of us are completely unaware of.

Plato didn't invent the idea that the Earth is round. And, while we cannot rule out that he found it in the book by Anaxagoras, it is more likely he picked it up from his friends among the Pythagoreans of Italy. However, before we label Plato as an unequivocal globalist, we need to note that elsewhere in his works he gave descriptions of the universe that are inconsistent with the Earth being a ball. For example, in a dialogue called the *Phaedrus*, he had another crack at revealing the fate of souls after death. He described the heavens as a vault held up by an arch, which the dead must ascend to reach the realm of the gods.[29] Although it isn't stated explicitly, this intimates at the traditional Homeric world picture of a flat earth with a dome over the top. But Plato was never too concerned with how well his descriptions of the universe corresponded to reality. And he failed to provide any evidence for the Globe even though it was a revolutionary hypothesis. It would be wrong to say he 'knew' the Earth is a sphere even if he happened to believe it. Knowledge requires us to marshal the reasoning that justifies our belief. Plato never tried to do that for the shape of the Earth. One of his students did.

8

Aristotle: 'Necessarily spherical'

You'll often hear it said that the ancient Greeks 'knew' that the Earth is a sphere. The last few chapters have shown that this wasn't the case prior to the fourth century BC. Like his contemporaries, even as outré a philosopher as Anaxagoras thought it was a disc. But why do we say 'I *know* the Earth is round, but Anaxagoras *thought* it was flat'? How come the shape of the Earth is knowledge for us, but Anaxagoras only gets to have a belief about it? Obviously, it's because we can't know something unless it's true. As the Earth isn't flat, you can't know that it is, any more than you can know spiders have ten legs. If you believe in decapod arachnids scuttling over a flat earth, you're wrong.

This raises a further question: what do we mean by truth? Most philosophers agree that a statement is true only when it corresponds to reality; in other words, it is true to say that the Earth is a sphere if and only if it is spherical. In philosophical argot, this is usually called the 'correspondence theory of truth'.

However, merely believing something that happens to be correct isn't enough to *know* it. For example, the astronomer Aristarchus of Samos (*c.* 310–*c.* 230 BC) said that the Earth orbits the Sun. Of course, he was dead right. But Aristarchus did not have good evidence for his hypothesis. For this reason, it would be wrong to say he knew how the Earth moves. His theory might have been true, but there was not enough data to justify it. A lucky guess doesn't constitute knowledge.

Plato defined knowledge in the *Theaetetus*, one of his dialogues, calling it 'true belief accompanied by a rational account'.[1] Today, epistemologists rephrase this by saying knowledge is a 'justified true belief' or 'JTB' for short.

The Babylonians and Egyptians, as well as the early Greeks, thought they knew that the Earth is flat. But they were mistaken. Although Plato believed it was a sphere, he would have readily admitted he didn't have enough justification to say that he 'knew' this. So, who was the first person to 'know' it? It was the one who gathered enough evidence to justify his or her belief that the Earth is spherical. That is, we need to find someone who not only believed that the Earth is round but had good reasons for doing so. I would contend that was the person who 'discovered' the theory of the Globe, just as we say Albert Einstein discovered the theory of Relativity.

Two parallel streams of investigation needed to coalesce to elevate 'the Globe' from a belief to knowledge. There had to be an accumulation of data from the nascent science of astronomy to provide empirical evidence of the Earth's shape. As we saw in the last chapter, Plato didn't feel obliged to provide anything like this. And someone had to develop a meaningful model of the cosmos in which a spherical earth could fit. This model needed to address objections to the Earth being a ball – above all, why we don't fall off. The combination of observations by astronomers and theories developed by philosophers combined into an inference that the spherical earth was the best available explanation of the evidence.[2] The person who pulled all the strings together and published the findings is our candidate for being the one who discovered the Globe, even if that was the culmination of work by many people over several decades.

Eudoxus and His Concentric Spheres

To witness the evidence for the Globe being collected, we need to visit Plato's Academy and meet some of his colleagues and students. They were the people who helped to justify a belief that the Earth is a sphere and turned this crazy idea into knowledge.

We don't have many details about what life was like at the Academy. The dialogue form of Plato's writings suggests that students and masters engaged in debates and discussions, but we can't be certain. We don't even know if Plato's published works represent what he was teaching. Luckily, there are hints in the dialogues that allow us to speculate about how Plato might have steered his pupils. In particular, there are tantalizing remarks about astronomy in the *Republic* and in Plato's unfinished final dialogue the *Laws*. These might refer to an astronomer and member of the Academy named Eudoxus of Cnidus (*c.* 400– *c.* 345 BC), who was responsible for gathering important evidence for the curvature of the Earth.

It's certain from the *Phaedo* that Plato was familiar with the concept of the Globe by the 370s BC, and that he could have picked this idea up from his Pythagorean contacts in southern Italy. By this time, Eudoxus of Cnidus had been part of Plato's circle for many years, spending time at the Academy in Athens and travelling widely. His home city was in the southwestern corner of Asia Minor, in modern Turkey. Like so many Greeks, he is said to have stayed in Egypt as a young man, learning the wisdom of the priests. While there, he was marked for greatness by one of the sacred bulls, which licked his cloak.[3]

Even if we disregard this as a myth, there is a good chance that Eudoxus did spend time on the Nile, since he commented on the river's flooding and carried out astronomical observations there. Later, he visited Cyzicus, not far from the ancient city of Troy, where he continued his investigations of the heavens. By the 360s BC, he was back in Athens and may have temporarily taken over leadership of the Academy while Plato was on another ill-fated excursion to Sicily. At some point, late in his life, Eudoxus returned to his home town of Cnidus, where he rewrote the local code of laws and thus earned the esteem of his compatriots. He died shortly after Plato, since he mentioned the philosopher's passing in one of his books.[4]

During his life, Eudoxus wrote about astronomy and geography, as well as mathematics. His most famous work was a catalogue of the constellations called the *Phenomena*. Although the original does not survive, a poet called Aratus (*c.* 315– *c.* 240 BC) rendered it into verse a few decades later. This poetic version, which takes up less than thirty pages in a modern paperback translation, became one of the most influential books of antiquity.[5] It fixed the names of the constellations that we use today and continued to serve as a guide to the skies as late as the sixteenth century. Aratus echoed the view of Anaxagoras and Plato that the universe is spherical, with the Earth at its centre. The stars fleck the inner surface of the sphere that bounds the cosmos, which rotates every 24 hours about an axis that runs through its north and south poles. Aratus didn't mention the shape of the Earth and continued to use traditional language from Homer about stars falling into the Ocean. This might be poetic licence, but it is just as likely that in the early third century BC, when Aratus was writing, most people still assumed the Earth was a disc.

Eudoxus carried out observations across a wide area of the eastern Mediterranean over the course of his career. For example, he spent time in north and south Anatolia, in Athens and in Egypt. This meant he was able to detect that the hours of daylight vary with latitude.[6] We all know that nights are longer in the winter, and days are longer in the summer. However, the closer you travel to the equator, the more equal the durations of darkness and daylight become. At the equator itself, they are both twelve hours all year. If the Earth were flat, we would expect the sun to be up at the same time for everyone.

Furthermore, Eudoxus found there were stars observable on the Nile that he couldn't see in Cyzicus or in Athens. The most obvious example is Canopus, the second-brightest star in the sky, which he called 'the star that rises in Egypt'.[7] You can't see Canopus from most of mainland Europe or the northern United States. It is visible for some of the year, low in the sky to the south, from Crete or Cyprus, and becomes very obvious when you get as far as Cairo. This is powerful evidence for the curvature of the Earth. From the Northern Hemisphere, you can see the stars that are over the North Pole, but not those below the Antarctic. As you head south, more of these southern stars come into view, while the northern ones disappear below the horizon. In the United Kingdom, the most obvious constellation is the Great Bear, which is almost directly over the Arctic Circle. As Homer noted, that means it is visible all year round. However, in much of the Southern Hemisphere, you can't see it at all. For Australians, the Southern Cross is the most familiar marker in the sky, but it is never seen from Europe or the United States.

The heavens appear to rotate around the Earth each day, but the constellations retain their shapes and positions relative to each other. Meanwhile, the five visible planets move in haphazard paths against the background of the fixed stars. Once Greeks like Eudoxus started to observe the planets closely, they saw that they travel across the sky roughly west to east at varying speeds. Sporadically, they even appear to turn back on themselves, a phenomenon called 'retrograde motion'.

We now know that the planets only appear to move backwards when observed from Earth because it and the other planets are all orbiting the Sun at different speeds. However, for the Greeks, retrograde motion was a puzzle. Plato found it downright offensive.[8] He thought the planets were divine and so all this random wandering around the sky was out of order. They should move around the Earth in circles because this was the perfect shape. In the *Republic*, he instructed his readers to leave off observations of the heavens until they could provide a theoretical explanation for what the planets were up to.[9] The challenge for the members of the Academy was to show how the apparent movements of the heavenly bodies could be reconciled to the requirement for uniform circular motion. Only in the seventeenth century did Johannes Kepler (1571–1630) figure out that the planets truly orbit the Sun in ellipses.

Eudoxus rose to the challenge Plato laid down in the *Republic*. According to Aristotle, he proposed a model of the universe made up of concentric spheres, starting from the outermost shell of the fixed stars.[10] Within this, each planet was mounted on a sphere that moved with the required uniform circular motion. To deal with the variations of speed and direction displayed by the planets, Eudoxus added further spheres. The overall path of the planets across the sky was a combination of their several movements. By adjusting their speed and direction of rotation, Eudoxus could approximate the observed motions of the planets in the sky.[11] Plato would have been thrilled about this. Indeed, in the *Laws*, his very last dialogue, he praised the astronomers who had rescued the divinity of the planets by synchronizing observation with the reality that they followed one fixed orbit.[12]

It is likely that Eudoxus assumed the Earth is a ball. That made sense given that all the planets sat on nested spheres that rotated around the Earth. Furthermore, the observations of the different stars that he could see in Greece, Asia Minor and Egypt had given him evidence for the curvature of the Earth's surface. Thus, by Plato's death, in around 348 BC, members of his Academy had valid reasons to believe the Earth is round. They could observe the curved shadow of the Earth against the Moon during a lunar eclipse. They knew from the observations of Eudoxus that the visible stars changed as they travelled south. And Eudoxus' model of concentric spheres, which appeared to explain the movements of the planets, made it logical to assume that the Earth itself was the innermost orb in the universe. However, the biggest problem had yet to be solved: why is it that we don't slide off the Earth's surface into the void? Without an answer to that question, no one could claim to know it is round.

The man who provided a solution was Aristotle: Plato's most celebrated student and the most important thinker in the Western intellectual tradition. Unlike Plato in the *Phaedo*, Aristotle was clear and unambiguous. Rather than just telling us what he thought, he set out reasoned arguments to justify his belief in the Globe.

Aristotle: The Philosopher

Aristotle's father was a physician well connected to the royal court of Macedonia. The family's wealth meant Aristotle never needed to worry about money and could concentrate on philosophy. As a young man of about twenty, he travelled to Athens and presented himself at Plato's Academy to complete his education. We don't know much about his time there, but we can assume he was a student of prodigious talent and ambition. Perhaps he found himself champing at the bit. Although he held his master in high esteem, his thought and methodology were radically different. In any case, by the time Plato died, Aristotle had already left Athens. He spent some time on the island of Lesbos before he was summoned back home to Macedonia, where he became tutor to the king's son, Alexander.

That Aristotle, the Western world's most significant philosopher, taught Alexander the Great, its most prolific conqueror, feels like it should be consequential. Sadly, nothing in the subsequent career of either party suggests they had much effect on each other. Aristotle hardly mentioned the Macedonians in his writing, even in the explicitly political books. As for Alexander, he was a brutal despot whose behaviour was as far from that of a philosopher king as it is possible to imagine.

Despite his distinguished teacher, it's unlikely that Alexander's education extended beyond familiarity with the epics of Homer, which any literate Greek was expected to know. As we've seen, Homer had thought the Earth was a flat disc with ocean running around its rim. All the ancient evidence, which admittedly dates from long after his death, shows Alexander shared this view. When his troops begged to turn back from India, he responded that he wanted to subjugate the ends of the Earth, all the way to the Ocean where the Sun rises.[13] He didn't think there was much further to go and, 'as the great stream of the Ocean encircles the Earth', it offered a convenient shortcut home.[14] Maybe Aristotle himself had not yet developed his theory about the Globe when he had to educate the Macedonian prince. Or he just didn't see a reason to mention a controversial idea that would have confused his young charge.

Returning to Athens in 335 BC, Aristotle founded his own school at a temple of Apollo called the Lyceum. Together with the rest of Greece, the city had become a tributary of Macedonia before Alexander marched off on his mission to subdue the Persian Empire. The philosopher was safe in Athens as long as it was under Macedonian domination, but in 323 BC, Alexander died in far-off Babylon and his vast realm began to fracture. Aristotle fled Athens to escape anti-Macedonian sentiment and a possible lynching. He died in exile a year later.[15]

His surviving works appear to be lecture notes from the Lyceum, possibly transcribed by his students. He is remarkable for the breadth of the topics he covered, including logic, ethics and natural history. Unlike Plato's dialogues, the prose in his books was not carefully polished for publication and is often frustratingly terse.

The Lyceum survived Aristotle's death but his reputation gradually waned over the next couple of centuries. In 44 BC, the Roman statesman Cicero (106–43 BC) noted he was 'not much known to philosophers, except to a very few'.[16] However, at around this time, his books were republished and his prestige began to grow afresh. His work would form the basis of advanced education throughout the Middle Ages in both Europe and the Islamic World, when he was simply called 'the Philosopher'.

Aristotle's Argument for the Globe

Aristotle constructed his cosmos using elements from the thought of the Presocratic philosophers who came before him. He held that the Earth is fixed, neither rotating nor moving, at the centre of an enormous universe. He had good reason to adopt this geocentric and geostationary model, not least because it was the consensus followed by pretty much everyone else. There was empirical evidence too. Look up on a clear night; over time, you can discern that the constellations appear slowly to rotate. Since we can't feel the Earth moving, it's natural to assume we are stationary and the heavens are turning. Also, if you throw a ball into the air, it falls straight down again, rather than getting left behind as the Earth turns beneath it.[17]

It was harder for Aristotle to explain *why* the Earth doesn't move. He understood that the basic issue was one we have considered before – the definition of downwards. When we drop a clod of dirt, it falls to our feet. Aristotle recognized that we instinctively believe that, if the Earth wasn't there, the clod would keep falling forever. And yet, the Earth appears to be immovable. Something is holding it in place. Aristotle rejected the explanations of earlier philosophers, such as the contentions of Thales and Anaxagoras, that it floated on water or was supported by a cushion of air. Unlike Plato, he was also unhappy with the suggestion of Anaximander of Miletus that the Earth stayed put because it was already at the centre of things.[18]

Aristotle's flash of genius was to redefine the meaning of 'down'.[19] Instead of it being perpendicular to the ground and running to infinity, he said it pointed towards the midpoint of the universe. Since he had already defined the universe as a huge but finite sphere, it had to have a centre. Drop something heavy, he said, and it falls towards this centre, unless something else gets in the way first. When it gets there, it stops. This explains why the Earth doesn't need anything to support it: since it's at the centre of the universe, it is already at rock bottom with no further to fall. So, it stays put. There was no need for Anaxagoras' air cushion or Thales' primeval waters for it to float on.

This did not mean to say that the Earth itself is the cause of gravity. Aristotle hadn't anticipated Sir Isaac Newton (1643–1727). He still had to explain why solid objects fall downwards. Here, Aristotle flounders a bit. His account begins with the idea that everything in the world – be it animal, vegetable or mineral – has behaviour that comes naturally to it. For example, human beings have an innate station in life and flourish when they occupy it. It just so happens that the highest calling for men is to become a philosopher like Aristotle, but he admitted that other people might be better suited as slaves and should accept that this is the place ordained for them.[20] He implied that a slave really would be happier in a state of bondage than being free.

Like Plato and the epic poets, Aristotle extended his ethical precepts to the natural world. He took what he thought was inherent in human society (but was actually just redolent of the culture he lived in) and applied it to the entire cosmos. Thus, he said, a heavy rock has an innate desire to move in a straight line towards the centre of the universe. So does water, just not quite so strongly, so it forms the next layer outwards above the Earth. Air is lighter, so the atmosphere sits over the waters. Finally, fire is the lightest element of all, so climbs through the air. This looks like an extension of Archelaus' idea that the cold elements of earth and water naturally sink, while hot air and fire tend to rise. Aristotle adapted his intuition that hot and cold elements

Aristotelian world picture, from Peter Apian, *Cosmographia* (1550 edn).

have a propensity to move in a particular way but recharacterized their properties as light or heavy for the purposes of explaining natural motion.[21] Illustrations of the Aristotelian world picture are immediately recognizable by their portrayal of the four elemental shells.

This solution was elegant. More importantly, by emphasizing that everything had its proper place, it ensured the rules that governed the natural world were consistent with Aristotle's ethical theories. However, it still required some extemporary components to explain what we see in the sky. He had said that the four elements tend to the centre of the universe. So why didn't the stars and planets fall out of the sky, like Anaxagoras' meteorite had done? Aristotle responded that the heavens consist of a fifth element, variously called 'aether' or 'quintessence'. This material resisted the lure of the centre and naturally orbited around it, rather than in a straight line towards it. The aether was not subject to corruption or entropy, so the stars track their frictionless trails forever. In contrast, beneath the orbit of the Moon, matter made of the other four elements was subject to change and decay.[22]

To explain the movements of the planets, Aristotle adapted the concentric spheres of Eudoxus (adding a few more for good measure). Beyond the bounds of the universe, he located his version of God. This deity didn't do much besides contemplate its perfection. However, this was enough to keep the whole universe in motion through the rotation of the outermost sphere. As it turned, its movement transferred to the sphere below it, and so on until it eventually reached the Moon, the lowest part of the heavens. For Aristotle, it was axiomatic that any moving object needed something else to move it, so an unmoved mover outside the universe had to make the world go round. Unlike Plato's divine craftsman, Aristotle's god wasn't a creator. The universe was everlasting – it had always existed and always would. The rotating celestial spheres marked time through eternity.

Aristotle had everything he needed to show that the Earth is a stationary sphere. He set out his analysis in *On the Heavens*. Typically, he would begin his enquiries on a subject by recapping what his predecessors had said. That was the context in which he recounted the cosmology of Philolaus, which we examined in the last chapter. Then, he sought axioms from everyday observation and common opinion that were accepted as true by everyone. Next, he used these axioms to infer general rules. Finally, he demonstrated the way that these general rules account for what we see happening.[23] Or, to know something is to explain its cause from basic axioms and to use it to explain other things.[24] Aristotle didn't follow this method for every subject he wrote about, but his elucidation on why the Earth must be round follows it quite closely.

In his review of earlier authorities, Aristotle noted that Anaximenes and Anaxagoras had thought the world was flat. However, he didn't record the name of anyone who said the Earth was a sphere. This indicates just how recent the idea was. In general, to be namechecked by Aristotle, you had to be dead. He didn't tend to quote or acknowledge his contemporaries. If he knew someone like Parmenides, who'd died a good century previously, had advocated for the Globe, Aristotle would probably have mentioned it. He did note that some unnamed people thought the Earth is spherical, but it is likely that he was referring to his own circle in Athens.

After analysing the views of his predecessors, Aristotle built out his world picture. As described above, he showed it was common sense that the Earth had to be stationary and at the middle of the universe. This was because all heavy things naturally fell centripetally. Finally, he could infer that the Earth itself must be a ball.[25] In their desire to reach their natural place, solid objects moved towards the heart of the heavens from all directions. This necessarily meant they coalesced into a sphere, which is the Earth. It doesn't even make a difference if heavy matter is predominantly falling from one direction as it will still jostle itself into the shape of an orb when it reaches the centre.

Aristotle concluded his discussion by showing how the theory of the Globe explained observations that might otherwise seem inexplicable. In the first place, he said that when there is a lunar eclipse, the shadow of the Earth on the Moon is always curved.[26] This corroborates what he had already shown from his first principles. The umbra during a lunar eclipse follows from the shape of the Earth. If it is a ball, its shadow must always be an arc.

His second piece of empirical evidence is the way the visible stars change as we travel north or south.[27] He noted that some stars, which are visible in Egypt and Cyprus, can't be seen in the north. He is almost certainly referring to Eudoxus' observation of Canopus. It is bright enough to be hard to miss in Egypt, albeit usually low in the sky. Its absence from view in Athens would have been obvious to anyone who had seen it further south. This is only explicable if the Earth is rounded since, if it were a flat plane, everyone would see the same stars. Since it is spherical, it's inevitable that our view of the heavens will change with latitude.

Aristotle observed you can use the elevation of the stars to estimate the size of the Earth, and also to deduce it must be tiny compared to the vastness of the heavens. The number he quoted for the circumference of the Earth, 74,000 kilometres (46,000 mi.), is almost twice the true value of 40,075 kilometres (24,901 mi.). It is likely that Eudoxus had estimated this figure by measuring differences in the height of the Sun or certain stars as he travelled south.[28] Aristotle was probably also referring to Eudoxus when he noted that it should be possible to travel westwards across the Atlantic to eventually reach India.[29]

Right for the Wrong Reasons

Aristotle thought he had proved that Earth is a sphere. He'd shown it ought to be round from first principles, and then used this conclusion to explain phenomena such as the shadow during a lunar eclipse and how the visible stars change as we travel around. But, by our lights, did Aristotle really *know*? Until recently, most philosophers would have said he did. His belief in the Globe was true, and he justified it both theoretically and empirically. Then, in 1963, a fledgling American thinker called Edmund Gettier (1927–2021) published a short article that argued it's quite possible to have a justified true belief in a proposition without also 'knowing' it, if your justification for believing it is awry.[30] Epistemologists found this a convincing rebuttal of Plato's definition of knowledge.

This spells trouble for Aristotle. He justified his belief in the Globe by explaining the Earth was the accumulation of all the solid material that had fallen towards the centre of the universe. This is false. Some of his ancillary evidence was dodgy as well. He noted that elephants lived both in the west of the known world, in Morocco, and in the far eastern part, in India. He supposed this might be evidence that the elephant populations joined up around the other side of the world. Yes, he buttressed his argument with observations of the Moon and stars, but these were not determinative by themselves. So, while Aristotle's argument for the Globe contains some shrewd arguments, it's based on completely false axioms about the structure of the cosmos. It is a fine example of getting something right for the wrong reasons.

Personally, I think we should be generous to Aristotle. His argument was hardly watertight but he got to the correct answer and used it to explain puzzling facts. The original concept may have been Pythagorean, and Plato had been discussing it at the Academy for several years while Aristotle was a student there. Eudoxus could have thought that the Earth was a sphere as a result of his astronomical and geographical studies. However, it was Aristotle who fitted the jigsaw together, combining the insights of this small group of people with the empirical evidence they had gathered. By any conventional standard, he knew the Earth was a sphere, and he was probably the first person who did. On that basis, he discovered the theory of the Globe. As we will see in the remainder of this book, everyone today who knows the Earth is round indirectly learnt it from Aristotle. This makes the Globe the greatest scientific achievement of antiquity. It's only because we take it as obvious that we don't give Aristotle the credit he deserves. I am as guilty as anyone, having written a book highlighting how he was wrong about almost everything else.[31]

However, we shouldn't imagine that Aristotle immediately settled the question of the Earth's shape. Even among Athenian philosophers, his idea wasn't unanimously accepted. Not all of them found it congenial to their world view.

9

Greek Debate on the Shape of the World: 'Either round or triangular or some other shape'

I f a time traveller offers you the chance of a guided tour of ancient Alexandria, be sure to take them up on it. Two thousand years of upheavals, erosion and war have left little trace of the original city, and its Royal Quarter has sunk into the sea. Alexandria sat athwart a knoll between Lake Mareotis (which has now largely dried up) and the Mediterranean, opposite the island of Pharos. A causeway linked the island to the mainland, dividing the harbour into two parts. A lighthouse, one of the seven wonders of the world, guarded the entrance to the eastern Great Harbour.

The streets were laid out in a grid aligned to the main thoroughfares, Canopic Avenue and Soma Avenue. At their intersection in the centre of the city, you would find the Soma itself, the tomb of Alexander the Great. His body was brought here when Ptolemy I Soter took over Egypt and made Alexandria his capital. The city was effectively a Greek colony that ruled over the indigenous Egyptians of the Nile Valley. North of Canopic Avenue stood the Royal Quarter containing the palace of the pharaoh. The famous Temple of the Muses or Museum was probably in the same part of town. Ptolemy's son and successor, also called Ptolemy – like every male member of the dynasty – encouraged writers and philosophers to make their home in Alexandria to bolster Greek culture and to increase his prestige as a patron of the arts. He financed the Museum to provide these immigrants with a home, and they repaid him by making the city the Mediterranean's foremost intellectual centre. Even today, we owe our knowledge of the text of Homer and aspects of classical Greek grammar to philologists who worked in Ptolemaic Alexandria. However, during your tour, you would have searched in vain for the Great Library. While the Museum undoubtedly had a formidable collection of books, the universal library, said to contain the entirety of Greek science and literature, is a myth, as are the various accounts of its destruction.[1] Our most detailed description of the ancient city, by the geographer Strabo (*c.* 64 BC–*c.* AD 21), makes no mention of it.

The Size of the Earth

Thanks to one Alexandrian scholar, the ancient and medieval world had a pretty good idea of how big the Earth is. Eratosthenes of Cyrene (*c.* 276–*c.* 195 BC) was a polymath who worked in the fields of geography, history, astronomy and pure mathematics. He was attached to the Museum and later tradition named him as one of the directors of the imaginary Great Library.

It's said that Eratosthenes pieced together an estimate for the Earth's circumference from Egyptian surveying records.[2] The surveyors would have measured the shadow cast by a vertical stick, called a gnomon, at noon on a specific day to determine the angle of the Sun's elevation in the sky. To work out the Earth's size, Eratosthenes needed these angles for at least two locations, one of which was due south of the other. He also needed to know the distance between them.

He chose Alexandria itself, the city of Syene (today's Aswan) in southern Egypt, and Meroë on the Nile in modern Sudan. He worked on the basis that the three cities were spaced

5,000 stades apart on a southern bearing. There's been a good deal of debate about how he came by this distance, but it is not a bad approximation. Finally, Eratosthenes calculated the differences in the angle of the Sun in Alexandria, Syene and Meroë. This turned out to be about one-fiftieth of a complete circle per 5,000 stades, which implied that the distance between Alexandria and Syene was one-fiftieth of the circumference of the Earth or 250,000 stades.[3] It is difficult to convert this into modern kilometres or miles because the length of a stade varied. It was supposed to be the size of the running track in a Greek athletics stadium, which was roughly 185 metres (200 yd). However, the stadiums in Greek cities were not all the same. For instance, the stadium at Olympia is slightly smaller than the one in Athens. In any case, it is likely that Eratosthenes' estimate of the size of the Globe was about 10 per cent too large. This is still a great deal more accurate than the figure of 400,000 stades recorded by Aristotle.

Eratosthenes didn't set out to prove that the Earth is a sphere. It was just one assumption that lay behind his calculations. Another was that the Sun is a very long way away. This meant that he could assume that the curvature of the Earth was entirely responsible for the difference in shadows between Alexandria and Syene.[4] By providing a reasoned estimate of the Earth's circumference, Eratosthenes helped to build a practical superstructure around the theory of the Globe. His figure of 250,000 stades would remain the standard for centuries to come, adopted by the Romans and taught at the universities of medieval Europe.

Eratosthenes shows that, in the third century BC, Greek scholars in Egypt, at least, knew the Earth was round. They were among the early adopters of the Globe. Nonetheless, from these small beginnings it gradually became accepted across the Greek world. Alexandrian astronomers did their bit to achieve this by trying to improve the model of the planetary motions developed by Eudoxus. For example, Hipparchus of Nicaea (*c.* 190–*c.* 120 BC) developed a more sophisticated picture of the heavens using a geometrical construction called an epicycle. Instead of the concentric spheres of Eudoxus, he placed the planets on small rotating circles that orbited the Earth on larger rings. This allowed him to reproduce the observed motions of the Sun and Moon more accurately than Eudoxus had done.

Ptolemy of Alexandria (*c.* AD 100–*c.* 170) developed an even more complicated model. We know next to nothing about his life, except that he was not related to the Alexandrian dynasty of the same name. His books represent the apotheosis of ancient Greek astronomy, astrology and geography.

At the beginning of his masterpiece, the *Mathematical Synthesis*, he presented evidence for the shape of the Earth that had not been available to Aristotle. He observed that the time the Sun rises varies as an observer travels east or west, so while it is dawn in Antioch it will still be dark in Rome. This, coupled with Aristotle's observation that different stars become visible as we travel north or south, showed the Earth curved in all directions, so it must be a sphere rather than a cylinder. He also mentioned the way that a mountain gradually appears over the horizon as we sail towards it – a rather obvious clue pointing to the Earth's curvature that wasn't noted by Aristotle.[5]

Improvements in Greek astronomy made it possible to accurately predict the paths of the planets and marvels like eclipses. As it became apparent that the theory of the Globe and a spherical universe could be used to explain things in the real world, the credibility of the model that undergirded it was enhanced. In parallel, it was being discussed by some of the philosophical schools that were proliferating in Athens.

New Schools of Philosophers

Few Greeks learnt about the universe from complex astronomical works by the likes of Hipparchus and Ptolemy. They were more likely to study philosophy, although not books by Aristotle, who'd fallen out of fashion in the decades after his death. Instead, the most influential philosophical school between 300 BC and AD 200 was Stoicism, and it was Stoic promotion of the Globe that led to its widespread acceptance, at least among the educated elite.

The Stoics' founder was a certain Zeno of Citium (c. 334–262 BC), who was allegedly washed up in Athens after being shipwrecked. Inspired by tales of the rectitude with which Socrates had faced his misfortunes, Zeno began teaching philosophy in the Athenian marketplace under a decorated portico called the stoa. This gave its name to the Stoic school.[6]

Stoics divided knowledge into three disciplines: ethics, physics and logic. Of the three, ethics was, in practice, the first among equals. Like other Greek philosophies, Stoicism gave its adherents a way to relate to the world so that they could enjoy a tranquil life, whatever disasters befell them. As with Platonism, the metaphysics and natural philosophy of the Stoics laid a foundation for their moral code. They lifted their picture of the physical world directly from Aristotle, agreeing with him that the universe comprised twin spheres: the larger encompassed the whole of existence, with the smaller – Earth – at its centre.[7] This picture fitted Stoic conceptions of the cosmos as a divine and eternal entity that went through cycles of destruction and rebirth.

We can't be sure if the Globe featured in the doctrine of Zeno himself, but it quickly became part of the package that all his followers accepted. Several Stoic textbooks on astronomy survive, including *Introduction to the Phenomena* by Geminus (*fl.* first century BC) and *The Heavens* by Cleomedes (*fl.* first century AD).[8] Neither author took the Globe for granted but laid out the astronomical evidence to convince their readers. It was probably books like these, used for teaching students, that spread the theory through the wider literate Greek population.

After the Romans annexed Greece in the second century BC, Stoic thought propagated among the conquerors' upper classes. High-profile Roman Stoics included Seneca (*c.* 4 BC–AD 65), the right-hand man of Emperor Nero (AD 37–68), and the emperor Marcus Aurelius (AD 121–180), both of whom wrote books on Stoicism that survive to the present day. As Greek philosophy spread through the ranks of educated Romans, the Globe became an accepted part of their world picture, as it had for the Greeks. That's not to say acknowledgement of the spherical earth was anything like ubiquitous. Even among philosophers, it was far from universally embraced. In particular, the school of Epicureans, named after their founder, Epicurus (341–270 BC), were resolute in their rejection.[9]

Back in the mid-fifth century BC, when Socrates was still a young man, a certain Leucippus made a far-reaching suggestion. Maybe, he said, the underlying structure of matter bore no resemblance to the reality that we perceive. Perhaps everything consists of particles floating around in a void, bumping into each other. These particles are the tiniest units of existence and so cannot be split into anything smaller. He called them 'atoms', meaning 'uncuttable' in Greek. Leucippus is a murky figure, even by the standards of the Presocratics. However, his ideas were embraced by Democritus (*c.* 460–*c.* 370 BC), a much more famous thinker. He was a wealthy man from northern Greece who travelled widely to expose himself to as many cultures as he could. He adopted the atomism of Leucippus and made it part of a materialist philosophy to support his humanist ethics.

Democritus was a very old man before Plato had written about the Globe, so it is no surprise that he assumed the Earth was flat.[10] More controversially, he claimed that the Earth is not unique: there are an immeasurable number of worlds of many shapes and sizes throughout the limitless universe. Aristotle found plenty to criticize in Leucippus and Democritus. For a start, he rejected the concept of atoms. He thought it was logically impossible for something to be uncuttable. It must, he said, always be possible to grind the fabric of reality more finely.[11] He also insisted that the universe was unique and of limited, if still vast, size.

Democritus doesn't appear to have founded a school, but his ideas were a major influence on Epicurus, whose name has entered the English language. We call someone an 'epicurean' if they enjoy the luxuries of life, especially fine cuisine. This is unfair on Epicurus. Like most Greek sages, he was concerned about how to achieve a state of contentment. His solution was to seek pleasure and avoid pain, a philosophy called hedonism. However, the Epicurean quest for pleasure did not involve indulging in food, sex and cheap thrills. Rather, Epicurus taught that contentment should be obtained through deeper pleasures, most importantly friendship and the absence of suffering. He held that the most serious form of suffering was the self-inflicted and unnecessary fear of things that we should not worry about. Thus, far from being the philosophy of the trencherman and paramour, Epicureanism was a way to achieve equanimity by learning not to be concerned with what we can't change. Instead, we should enjoy the company of others, preferably in a pleasant spot like a garden. Our sources agree that hanging out with Epicurus himself would have been a joy. He was a thoroughly good egg, esteemed by his countrymen and a friend to all mankind.[12]

We possess substantial fragments of Epicurus' own writing. Three of his letters were reproduced in full by Diogenes Laertius, while bits of his major book *On Nature* were preserved as a result of the eruption of Vesuvius that buried Pompeii in AD 79.

Before the main eruption, the volcano had been scattering ash and pumice over the city for several days. This performance was just the volcano clearing its throat. The main explosion blasted ash and superheated gas high into the air. When this dark column of debris collapsed back onto itself, it rolled down the sides of the mountain, incinerating everything in its path. For centuries, the description of a young eyewitness who had seen the eruption was dismissed as hyperbole. He wrote of a 'fearful black cloud rent by forked and quivering bursts of flame', which then sank back down to the ground and rolled out into the Bay of Naples.[13] Vulcanologists today recognize this spectacle as a pyroclastic flow. The hot cloud of ash from the volcano also swept over the nearby coastal town of Herculaneum, cremating the population as they sheltered by the sea awaiting rescue. The intense heat burned thousands of papyrus scrolls held in the library of an enormous villa on the shore. They were not reduced to ash but turned into stumpy sticks of charcoal. In the eighteenth century, early archaeologists exhumed the scrolls from the ruins of the villa, which had been buried under the lava that enveloped Herculaneum in the aftermath of the eruption. Over the years, there have been many attempts to read their contents. Early efforts to unroll them ruined some beyond repair. Today, scientists use non-invasive methods such as X-ray phase-contrast tomography to read the scrolls without having to attack them physically.[14] From what it has been possible to extract in the last couple of centuries, it is clear the villa of the papyri held a library featuring works by Epicurus and his followers. Combined with the material preserved in the manuscript tradition, this means we can glean a great deal about ancient Epicurean thought.

To help practitioners of his philosophy achieve contentment, Epicurus taught how to achieve a state of tranquillity by banishing fear – primarily the fear of death and the gods. To help conquer it, he co-opted the atoms of Democritus. His materialist conception of an eternal and untrammelled universe nullified questions about who made the world or what lay outside it. Unlike Plato, he had no call for a creator. Not that Epicurus rejected the existence of the gods altogether. He didn't need to be that controversial. He maintained that their perfect immortal existence was such that they didn't bother with human beings, let alone take time out from their blessed state to start chucking around thunderbolts or to sniff the aroma from the sacrifices in the temples.

Having dismissed any concerns about divine agents, Epicurus had to deal with the problem of death. Since everything is just atoms and void, it follows that souls must be made up of minute particles as well. Epicureans said that the atoms of the psyche were especially fine such that, once our bodies died, they dissipated into the atmosphere. Thus, death automatically led to the extinction of consciousness. Since we will not experience being dead, and certainly don't have to worry about punishments beyond the grave, fear of our demise is, according to Epicurus, irrational.[15]

Like his predecessor Democritus, Epicurus rejected the Globe. He had good reasons for this. For instance, Aristotle's theory depended on the Earth being at the centre of the universe. If the universe had a central point, that implied that it also had an edge. This was a problem because it was inherently improbable that the random movements of atoms would form into things like giraffes, grapes and emeralds. Rather than open this can of worms, Epicurus made the universe limitless and everlasting. The bubble we live in is one of many worlds, which could be utterly alien in their appearance. They might be round, or some other shape, maybe even triangular.[16] We know from his surviving letters that he also thought that the Sun and Moon were rather small and nearby.[17] By downgrading the heavenly bodies to mere atmospheric phenomena, he stripped them of their divinity and showed that it was pointless to worship them.

As with Aristotle and Plato, the world picture of Epicurus buttressed his ethics. But as Epicurean hedonism is so very different from Aristotle's ethics, he rejected Aristotelian cosmology as well. At times, his opposition to Aristotle went beyond the professional and became deeply personal. He accused him of being a wastrel who squandered his inheritance and trafficked drugs.[18]

Our knowledge about Epicurus' natural philosophy has been further fleshed out by a Latin poem written by a Roman enthusiast for his teaching. It survives in its entirety, copied by ancient scribes and medieval monks until it became widely known in the fifteenth century. It's called *On the Nature of Things* by Lucretius.

The poem is dedicated to a certain Memmius, usually identified as the Roman senator Gaius Memmius (d. c. 49 BC).[19] After being accused of electoral fraud in Rome, Memmius retired to Athens in about 52 BC, where he wrote erotic poems. On arrival, he bought a plot of land on which he planned to build a house. According to Cicero, the land contained the ruins of the school founded by Epicurus, which Memmius wanted to demolish to provide space for his new villa.[20] At the time, Athens was far from being the philosophical centre it had been in previous centuries. In 88 BC, the Roman general Sulla (138–79 BC) had crushed a revolt in Greece and closed the philosophical schools on suspicion of political sedition. Still, within a few decades, philosophers were beginning to return, so wealthy Romans like Cicero would send their sons to Greece to acquire an education.

It's possible that the property development by Memmius provided the occasion for the dedication of *On the Nature of Things*. The Epicureans of Athens were outraged that the site of their founder's school was going to be turned into the retirement home for an amateur pornographer. Lucretius had been working on his poem for years but may have inserted Memmius as its addressee to convince the boorish Roman that he was obliterating a site of importance.

Expressing the doctrines of a subtle philosophical system in Latin rather than Greek was no easy task. Lucretius found that words for many concepts simply didn't exist in his own language. However, he was not an original thinker and followed the dogma of Epicurus slavishly. Indeed, his adherence to the original doctrines of his master and lack of interest in contemporary science and philosophy has seen him called a 'fundamentalist'.[21] He interpreted Epicurus as proposing that the world was rather like what we would call a snow globe. We all live on a flat surface covered by a rounded vault, within which the stars and planets move around like the flakes of white when the snow globe is shaken. The whole contraption is falling through the void and will eventually decay into its constituent atoms. The infinite space of the universe contains countless other worlds, each of which contain their own ground and sky.

Like Epicurus, Lucretius tried to convince readers that the heavenly bodies were not very big or far away, and that they can even be buffeted about by the winds.[22] And he declared that it was daft to imagine that there could be animals living on the other side of the world because they would inevitably fall into space.[23] Without Aristotle's explanation that matter tends towards the centre of the universe, the Globe made no sense. Naturally, Lucretius was ignoring the work of more recent astronomers like Hipparchus, who knew the Earth was a sphere, the stars were distant and the Sun was much larger than the Earth.[24] Indeed, by the first century BC, when Lucretius was writing, Epicurean cosmology had become quite an eccentric view among intellectuals. Cicero ridiculed the idea that there might be a multiplicity of worlds with different shapes.[25]

All this is a bit puzzling if we treat Lucretius as some kind of proto-scientist, as he is often presented today.[26] But 'science' was not what he was about. His purpose, like that of Epicurus, was ethical and everything he said about the physical world reinforced that message. He wanted to convince us that if we can grasp how the world really works, we will no longer be afraid. Only once we have escaped from fear can we be truly content. It doesn't even matter if our beliefs about nature are false. Epicurus was explicit on this point: a completely accurate physical theory was beside the point.[27] As long as there was a plausible naturalistic explanation for an experience, the wise man had nothing to fear. Aristotle's universe got in the way of that, so Epicurus and Lucretius rejected it.

The Epicureans were a well-known philosophical school but they never penetrated Roman society to the extent that the Stoics did. By the fourth century, they were all but extinct. Julian, the last pagan emperor, noted that most of the works of Epicurus had already disappeared, something he was not displeased about.[28] Lucretius' poem *On the Nature of Things* only survived because later Christian copyists thought it was worth preserving.

Greek Maps of the World

Greeks and Romans who accepted the Globe could live with Epicureans disagreeing with them. The primitive world picture of Homer was more of a problem. His reputation far exceeded that of a mere poet. The *Iliad* and *Odyssey* were as close to Holy Writ as the Greeks possessed, so people assumed that what he wrote had to be true. This led some ancient scholars to ignore the plain words of the text and deny that Homer thought the Earth was flat. A prime offender was the grammarian Crates of Mallus. Although he specialized in Greek, he is credited with launching the study of Latin grammar as well. It's said that, on a diplomatic visit to Rome, he broke his leg by falling into a sewer. During his enforced convalescence, he gave lectures that led the locals to imitate his methods of literary criticism.[29]

Crates was notorious for reinterpreting the *Iliad* to make out it described a spherical earth. This exasperated Geminus of Rhodes, one of the Stoic writers of astronomical textbooks. He rubbished Crates' anachronistic analysis of the epic, observing that no one in Homer's time had thought the Earth was anything but flat, the great bard included.[30] As for Crates, he wasn't above amending the Homeric text so that it cohered more closely to his theses.[31] He built a model globe, some 3 metres (10 ft) across, showing four continents, each separated by a broad sea. Tracing the wanderings of Odysseus around these landmasses, he asserted that, contrary to popular belief, the voyage in the *Odyssey* had encompassed the whole world rather than being confined to the Mediterranean Sea.[32] As we'll see, Crates' vision of the four landmasses would enjoy a long afterlife.

We don't hear much about terrestrial globes in the ancient world, but the Greeks had plenty of maps. According to tradition, the Milesian philosopher Anaximander drafted the first one. This was almost certainly circular, placing the Mediterranean Sea in the centre of the world, with the three continents – Europe, Asia and Africa – arranged around it.[33] The Mediterranean still bears a name that alludes to this, as it means 'middle of the earth'. The Ocean ran around the outside of these continents, just as Homer had said. The historian Herodotus, writing in the mid-fifth century BC, castigated the cartographers of his time who based their work on Anaximander's archetype. He had travelled widely and developed firm views about the dimensions and shape of the continents, insisting they were not of equal size and arranged in a rectangle rather than a circle.[34] That said, Herodotus had no idea of the Globe and couldn't even get his head around the fact that the Sun appears in the northern sky from southern Africa.[35]

Aristotle agreed that round maps were ridiculous and that the inhabited world had to be contained within a strip around the side of the Earth between the Tropic of Cancer and the Arctic Circle.[36] Today, we think of the Arctic and Antarctic Circles and the Tropics of Cancer and Capricorn as being lines on the Earth's surface. However, they started off as circles on the celestial sphere. Greek astronomers circumscribed the area of the sky in which the Sun moved over the course of the year with the Tropic of Cancer and Tropic of Capricorn, each of which was about 24 degrees away from the celestial equator. The word 'tropics' comes from the Greek for 'turning', since the Sun changed direction and headed back towards the equator when it reached them. This means

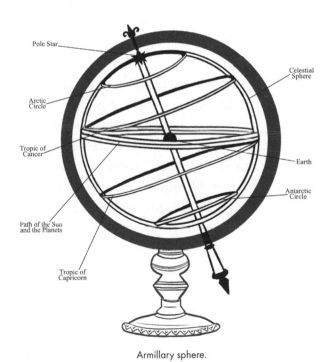

Armillary sphere.

that if you are standing on the Tropic of Cancer or Capricorn on Midsummer's Day, the Sun will be directly overhead at noon. The Greeks defined the Arctic Circle as being the same distance from the North Pole as the Tropics of Cancer and Capricorn are from the equator.[37]

Aristotle had the bright idea of projecting these heavenly circles onto the Earth, defining the area on each side of the equator as the torrid zone and that within the Arctic and Antarctic Circles as the frozen zone. The inhabited world necessarily had to be sandwiched between them, since the extreme climates of both the tropics and the ice cap rendered them unable to support human settlement.[38] An interesting implication of this theory was that there was another habitable region in the Southern Hemisphere, although whether anyone lived there is a ticklish question.[39] Aristotle's system of five climatic zones lasted until the sixteenth century, while the circular maps he so vehemently criticized became the most common image of the inhabited world throughout the Middle Ages.

Once Greek cartographers knew the Earth was a sphere, they needed a way to project a map of its curved surface onto a flat sheet of paper. It might seem sensible to try representing the surface of a sphere as a circle. This serves quite well when you want to map the heavens because a celestial chart simply shows what we can see. It doesn't need to make any allowance for stars being at vastly different distances from us. This doesn't work for a map of the Earth. You need good knowledge of geometry to project any large portion of the Earth's surface onto the page. And no matter how you do it, you'll distort distances or areas. The key is to use a projection that's consistent and maximizes the usefulness of the map. The most sophisticated ancient analysis of map projections was composed by the astronomer Ptolemy of Alexandria in his *Geography*, which featured latitudes and longitudes for thousands of locations.

Map of the climatic zones, illumination from William of Conches, *Philosophia mundi* (1277 or after).

Although none of his original maps have survived, the co-ordinates he listed allowed later cartographers to reproduce his depictions of the Earth and many of its regions. His map of the inhabited world shows Europe, Africa and Asia on a great curved apron around the North Pole.

Naturally, Ptolemy made mistakes. For instance, he believed the world to be a good deal smaller than it really is.[40] He also enclosed the Indian Ocean within land, making it impossible to sail from Europe to India. As we'll see, both of these issues would prove significant when Ptolemy's *Geography* was rediscovered during the Renaissance. That said, Ptolemaic maps of Europe and the Middle East are recognizable and show that ancient Greeks did have a reasonable idea about what the western end of Eurasia looks like. They were also aware of the vastness of Asia to the east, about which they knew very little.

Each time someone like Ptolemy used the Globe to produce a predictive model of the heavens or a map of the world, the evidence for the truth of the theory grew. And once the Aristotelian world picture had been incorporated into other philosophical systems, it became firmly entrenched among the literate elite. However, the dissemination of the Globe beyond Greek-speaking lands required more violent means: the conquest of the Mediterranean basin by the Romans. Their empire stretched from the borders of Scotland to the deserts of Mesopotamia, and they spread elements of Greek learning to ordinary people and not just the Stoic gentry.

10

Romans on the Globe: 'The circle of the world'

The Romans seem to have been strangely careless with their money, dropping it all over the place. From priceless hordes of golden *solidi* to a single worn bronze bit, amateur and professional archaeologists have recovered thousands upon thousands of coins. Currency was ubiquitous in the Roman Empire. It allowed trade to flow freely, while smaller denominations were essential to maintain a cash economy at the local level. Minting coins was a prerogative of the emperor, as it gave him a medium on which to communicate with his subjects. Today, the countless surviving examples are invaluable evidence of the kinds of symbolism that were supposed to be meaningful to ordinary people.

The obverse side of most coins was stamped with a bust of the emperor in profile, while the reverse allowed for more creative iconography. Spheres and orbs were common motifs. For example, the emperor Domitian (AD 51–96) showed his infant son sitting astride a globe surrounded by seven stars – most likely the seven planets that orbited the Earth.[1]

Let's look at a single example of a Roman coin, which I have beside me as I write. It's called a 'follis', a bronze disc less than an inch across that would have been loose change when it circulated. Everything we know about the coin is contained in its inscription. It was minted at Trier, now on the western border of Germany, under Emperor Constantine (*c.* AD 280–337) in

Roman coin of Emperor Constantine, Trier, AD 317,
with his patron god, Sol Invictus, holding Earth on the reverse.

AD 317. His head appears on one side, while the reverse portrays his patron god, Sol Invictus or 'the unconquered sun'. In his left hand, the sun god holds the Earth. Although he was the first Christian emperor, Constantine started his career as a pagan and never fully repudiated the traditional aspects of Roman public life.

It's unlikely that everyone who used the coin would have recognized the orb clutched by the sun god as the Earth, but the ubiquity of globes on Roman currency from the third century AD onwards shows that, by this stage, they were familiar to ordinary people.[2] How knowledge of the Earth's shape spread through the population is unclear, but it was probably a gradual process. It's not as if everyone was taught about it at school, as most children received no formal education.

Cicero's Dream

Today, we indelibly associate astronomy and philosophy with ancient Greece, but few people at the time had an appreciable interest in these subjects. The most technical books, like Ptolemy's *Mathematical Synthesis*, were rare and read only by specialists. That so much Greek scientific writing is available today reflects the interests of medieval Christian and Muslim copyists rather than its prevalence in the classical world. Likewise, philosophy was a consciously elitist pursuit. The contempt with which Aristophanes portrayed Socrates in *The Clouds* is a good indication of how commoners perceived philosophers.

We can get an idea of the books ordinary people owned from the scraps of papyrus archaeologists have extracted from ancient Egyptian rubbish dumps. Papyrus scrolls are made with criss-crossed strips of a reed-like plant. Unlike the clay tablets of the Mesopotamians, papyrus is perishable but the dry climate of Egypt helped to preserve it.

Much of the writing on papyrus is in Greek since that was the language of literate people in the eastern Mediterranean. Of the thousands of fragments from literary works (as opposed to letters and administrative documents), a full one-third come from the works of Homer.[3] There's hardly any theoretical astronomy, and when this subject does appear, it's almost always in the form of astrological almanacs and tables.[4] Ptolemy's *Handy Tables,* used by astrologers as a crib sheet to calculate the positions of the planets for their prognostications, turns up several times but not his more advanced works.[5] The only popular book of descriptive astronomy was the *Phenomena* of Aratus, which doesn't explicitly mention the shape of the Earth and implies it is flat. The least unusual source that mentions the Globe was Plato's *Phaedo.*

Egypt came under the rule of Rome when its last queen, Cleopatra VII (69–30 BC), picked the wrong side in the civil war between Mark Antony (83–30 BC) and Octavian Caesar (63 BC–AD 14), who later assumed the title of Emperor Augustus. Augustus defeated Cleopatra and annexed Egypt, uniting almost the entire Greek-speaking world, from Athens to Alexandria, within the Roman Empire. The conquerors were awed by the cultural and artistic achievements of the people they had sub-jugated. Upwardly mobile Romans sent their children to be educated in Athens, bought or pillaged all the Greek art they could lay their hands on and patronized Stoic and Epicurean philosophers. As the poet Horace noted, 'Captive Greece took captive her savage conqueror and brought the arts to rustic Latium.'[6]

Few Romans did more to educate his countrymen about Greek philosophy than Marcus Tullius Cicero. Cicero was a successful politician, the first of his family to become consul, who campaigned against the dictatorial tendencies of gener-als like Mark Antony and Julius Caesar. He was murdered during the civil wars that broke out in the aftermath of Caesar's assassination.

While Cicero was out of favour in Rome, he retired to his villa and wrote dialogues to expound the doctrines of the Platonists and Stoics, as well as castigate the Epicureans, of whom he did not approve. His substantial oeuvre, much of which survives, remains an important, if second-hand, source of information about Greek thought.

Cicero accepted the Globe simply because that was what educated Greeks believed. The fullest account of his world picture is in *The Dream of Scipio*, a fantasy that formed the con-clusion to his book of political philosophy called the *Republic*. The Scipio of the title was a real Roman general who is visited in a dream by his dead grandfather (also called Scipio and also a real Roman general). Scipio Senior takes his grandson on a voyage into the heavens so he can see that mighty Rome's con-quests cover just a fraction of the Earth's surface, while the Earth itself is but a speck in the vastness of the cosmos.

Babylonians and Egyptians had built their world pictures around themselves, imagining that Babylon was at the mid-point of the Earth or that the Nile bisected it. Cicero realized this wasn't possible if the Earth was a sphere. He couldn't maintain that Rome was at the centre of anything. However, he still found an ethical message in his vision of a universe where everything faded into insignificance. It was a lesson in humility not hubris. Carl Sagan (1934–1996) made the same point when he observed that Earth as seen from Saturn was just a 'pale blue dot'.[7]

In passing, *The Dream of Scipio* also imparts a good deal of geography to its readers. It describes the Earth as hosting four great continents separated by uncharted oceans, an account ultimately derived from Crates of Mallus. Cicero also covered the theory of the climatic zones, explaining that the heat of the tropics cut off the Southern Hemisphere from the north. Likewise, the Arctic and Antarctic were too cold for habitation, leaving just the temperate regions for people to live in. Even these areas were largely ocean or wasteland.[8]

The Romans called the world that they were busy conquering *orbis terrae,* which meant the 'circle of the earth'. Cicero clearly stated that *orbis* means a circle and not a sphere, for which the word is *globus.* He even provided the Greek translations of these words.[9] Traditionally, the Romans had subscribed to the world as a flat disc but as it became widely accepted that the Earth was spherical, the word *orbis* came to mean 'round' more generally. We can't tell exactly when the phrase *orbis terrae* became a reference to a sphere rather than a disc. It's likely both concepts existed simultaneously. The ambiguity around *orbis* has caused endless trouble for translators of Latin just as it has with Greek. It is never safe to assume that the word means spherical rather than circular, so, without proper context, we can't be sure when a Latin text uses the term to refer to a ball or a platter.[10] We'll come across instances of this confusion later.

Pliny and the Romans' Views on the Shape of Earth

Despite the ambiguity of their vocabulary, it's clear that, by the first century AD, literate Romans understood that Earth is a sphere. It's more difficult to establish what common folk thought, not least because the intelligentsia had little interest in them. Even the rubbish dumps of Egypt can only tell us about the reading habits of people who could actually read. One author who does mention the opinions of those he calls 'the vulgar' is Pliny the Elder (AD 23–79), who died during the eruption of Vesuvius. He was the uncle of the boy who witnessed the pyroclastic cloud that enveloped Pompeii. Both of them were staying in a nearby villa when Pliny the Elder left to take a boat across the bay to see if he could help with the evacuation. Despite being several miles south of Vesuvius, he died from asphyxiation after getting caught in the ash cloud.[11]

The Elder Pliny was a Roman aristocrat who wrote in Latin for his peers among the upper crust. His *Natural History*, an enormous encyclopaedia in 37 volumes, survives in full. It is a smorgasbord of facts and fantasies that its author had gathered from reading a plethora of books. It's fair to say that Pliny didn't quite master all the material he had browsed and often garbled his exposition. Even so, his book does give a good indication of the level of knowledge that would have been available to an erudite Roman.

In the section of *Natural History* on astronomy, Pliny noted that there was a consensus that the Earth was a sphere.[12] Unfortunately, after that, things get more confused. He castigated the common folk for imagining that people on the other side of the world ought to fall off and that the oceans should be flat. They can't understand, he said, how water can form a sphere around the Earth by sinking as close to the centre of the universe as it can. Pliny showed the Ocean is not flat by noting how a ship sinks below the horizon as it drifts away from the shore, and how you can see further from the top of the mast than you can from the deck.[13] In all, he set out more evidence for the curvature of the Earth than any other ancient author, mentioning the variable visibility of stars at different latitudes, including Canopus, and that the Globe explains why the hours of daylight differ across the world.[14] Most impressively, he explained how the signal carried by a series of fire beacons can travel so quickly that it becomes possible to see that the time of day changes from east to west, much as air travel gives us direct experience of the same effect.[15]

Despite the amount of material Pliny gathered, it's fair to say that precision was not his forte. Some of his incomprehensible explanations desperately need an editor. For example, he used the words *orbis* and *globus* in the same sentence without any indication that they might mean different things. At one point, out of the blue, he suggested the Earth could be shaped like a pine cone.[16] We are left with a tantalizing hint that the debate about the shape of the Earth among educated Romans had more nuance than just 'round' versus 'flat'.

The Globe in Roman Literature and Education

We find more traces of uncertainty about the Earth's form in some of the poetry composed around this time. One example is the *Georgics* by Virgil (70–19 BC). Unlike the *Aeneid*, Virgil's militant epic on the foundation of Rome, the *Georgics* celebrates the struggles and achievements of agriculture. Near the beginning, there's a primer on the constitution of the universe providing context for farmers' obsession with the weather.

Five zones comprise the heavens; whereof one is ever glowing with the flashing sun, ever scorched by his flames. Round this, at the world's ends, two stretch darling to right and left, set fast in ice and black storms. Between these and the idle zone, two by grace of the gods have been vouchsafed to feeble mortals; and a path is cut between the two, wherein the slanting array of the Signs may turn. As our world rises steep to Scythia and the Riphaean crags, so it slopes downward to Libya's southland. One pole is ever high above us, while the other, beneath our feet, is seen of black Styx and shades infernal.[17]

Virgil was certainly aware of the five climatic zones. He noted how two habitable regions were sandwiched between frozen wastes and the deserted tropics. However, he hadn't let go of the idea that there are absolute directions of up and down, so he placed the North Pole above us and the South Pole below. Given the Antarctic is beneath our feet, he also associated it with the underworld. While this might be poetic licence or an attempt to depict Greek cosmology in everyday language, the poem strongly implies Virgil knew about the Globe but didn't quite grasp its implications.[18]

Ovid (43 BC–AD 19), a near contemporary of Virgil, is best known for his bawdy love poetry, which might have prompted Emperor Augustus to exile him from Rome. He also composed a compendium of Greek mythology called *Metamorphoses*. In his retelling of the creation of the world from chaos, Ovid appears to refer to the Aristotelian universe, where a round earth is at the centre of concentric shells of water, air and fire. It's vexing that he uses the word *orbis* to describe the Earth, so we cannot be totally confident about whether he is calling it a disc or a sphere.[19] Still, it is likely that he knew about, and expected his readers to understand, the Aristotelian world picture. The same seems to be true of Lucan (AD 39–65), a poet who was a favourite of Nero before falling out with the emperor and being forced to commit suicide aged just 25. In his epic about Julius Caesar's civil war, Lucan told of soldiers lost in the desert who ask if they have wandered so far that Rome is now beneath their feet.[20]

In summary, knowledge of the Globe percolated through the rungs of society via a variety of sources including popular literature, basic education and visual media such as coins. This would not have been a quick process and was incomplete at the time the Roman Empire began its decline and fall in the fifth century AD. In particular, people from the western provinces such as Gaul and Britannia, where Greek influence was less pervasive, would have been slower to absorb these ideas.

Neoplatonism and the Globe

By AD 300, Stoicism and Epicureanism had lost much of their lustre. The dominant philosophy among intellectuals in the Roman Empire, even as, religiously, it turned Christian, was Platonism. This was not simply the philosophy of Plato reheated. A great teacher called Plotinus (AD 205–270) had developed a more mystical version of Plato's thought in the third century, which today we call Neoplatonism.

The Neoplatonists were content with the Aristotelian model of the universe, including the Globe. They developed a hierarchical world picture that consisted of a great chain of being from inanimate objects at the bottom, through spiritual beings and up to the godlike 'One' at the highest level. The Earth, at the centre of the universe, was not in a privileged position. Rather, it occupied the dustbin of the universe, to which base matter fell and from which good souls sought to escape.

The old Stoic astronomy textbooks by Cleomedes and Geminus fell from fashion, to be replaced by a new generation of manuals that described the same world picture in Neoplatonic terms. Several of these textbooks were composed in Latin to provide Roman students with a wide-ranging, if rather basic, philosophical education. These survived the fall of the western Roman Empire and formed the backbone of secular learning during the early Middle Ages.

A notable example is a detailed commentary of Cicero's *The Dream of Scipio*, written in the early fifth century by one Macrobius (*fl. c.* AD 400). Unperturbed that his brand of Platonism was nothing like Cicero's, he fleshed out the details of the *Dream*, providing background from Greek scientific authors and hammering home the ethical precepts that he believed undergirded the universe. He taught that the Earth is merely the size of a point compared to the diameter of the Sun's orbit, but he also recorded the circumference of the Earth as 252,000 stades, which he attributed to Eratosthenes.[21] This provided scientific ballast for Cicero's ethical point about the insignificance of even the greatest of men next to the vastness of the universe. By adopting Cicero's world picture, Macrobius transmitted the conjecture of Crates of Mallus that there are four great continents separated by uncharted ocean.

Writing around the same time, a certain Martianus Capella (*fl. c.* AD 420) structured his textbook around a marriage between the god Mercury and a maiden called Philology, who represents learning. At the wedding, each of the seven liberal arts, personified by a cultured young lady, gives a speech about her field. The liberal arts were so-called because they were the disciplines the Romans thought it fit for a free man to study. They remained the backbone of education right the way through the Middle Ages. The seven were divided between the trivium of grammar, logic and rhetoric; and the quadrivium of arithmetic, geometry, astronomy and music. Students tackled the trivium first, which gives us the word 'trivial'. The quadrivium was more challenging, and many pupils contented themselves with just a smattering of mathematical knowledge.

Martianus Capella covered the Globe in a brief section on geometry, which literally means 'measurement of the Earth'. 'The shape of the Earth is not flat, as some have supposed, who imagine it to be like an expanded disc,' he explained, 'nor is it concave as others suppose.'[22] Martianus needed to set out multiple arguments for the Globe because he knew it was counter-intuitive. As proof, he presented in evidence that the visible stars vary with latitude and that eclipses don't occur at the same hour of the day all over the world. He also noted that sundials need to be adjusted depending on where they are being used because the height of the Sun differs from place to place. Unfortunately, because Martianus, like Pliny, was relying on multiple sources, the information he gave wasn't always consistent. At one point, he echoed the consensus that the Earth is 252,000 stades round, while later on he gave an inflated estimate of 404,000 stades, more in line with the figure quoted by Aristotle.

By the time Martianus Capella and Macrobius wrote their textbooks, over seven centuries had passed since Aristotle first presented his case for the Globe. Since then, his world picture had escaped from highfalutin philosophical lectures into literature and poetry, and even imperial iconography. Educated Greeks and Romans knew the Earth was a sphere, had access to good arguments for its shape and had a rough idea of its size. Even many illiterate commoners were vaguely aware of these ideas.

News of the Globe was carried beyond the Roman Empire by travellers, scholars and merchants. The first country from where we have indisputable evidence of it becoming widely accepted among astronomers is India.

11

India: 'The mountain at the North Pole'

He might have been the most accomplished Indian astronomer of his day, but Lalla (c. AD 720–c. 790) had good reason to be grumpy. He'd mastered mathematics, astronomy and astrology, and his textbooks on these subjects set out everything needed to calculate the correct times to perform the sacrifices demanded of the Brahmins. And he had used his expertise to update venerable astronomical tables to make them current for his own era of AD 748.[1] But plenty of people were getting things wrong. 'Some say', he complained, 'that the Earth is supported by a turtle, a serpent, a boar, an elephant or by mountain ranges.' This obviously could not be. 'If the Earth is supported by a turtle or other things, by what are these things supported in space?' he demanded. 'If they can remain in space [unsupported], what prevents the Earth from remaining in position likewise?'[2] In other words, it can't be turtles all the way down.[3] Just as irritating were the people who insisted the world is flat. 'If the Earth is level, why cannot tall trees, like the date palm, be seen by man when they are at a very great distance from the observers?'

The People of the Veda

The multifarious beliefs about turtles and elephants criticized by Lalla come from a collection of books called the Puranas.

Most of them are written in an Indo-European language called Sanskrit, introduced into India by Aryans, who had migrated from the northwest before 1000 BC. As well as speaking an early form of Sanskrit, they rode in chariots and herded cattle. Like many pastoralist invaders, they installed themselves as rulers over the indigenous population.

We know something about the early Aryans of India because of the Sanskrit texts they left behind. The most venerable is the Rig Veda, composed in the area of modern Pakistan and handed down orally over many centuries. It is a compilation of hymns to the gods of the invading tribes. The archaic language has allowed scholars to date it to around 1000 BC, even though the surviving manuscripts are much more recent. The Vedas, of which the Rig Veda is the most ancient and revered, are now the most important scriptures for modern Hindus. Hinduism has developed over the centuries and gradually spread south across the Indian subcontinent, even to those parts to which the Aryans themselves had rarely ventured.

The Vedas reflect a warrior culture devoted to horses and cows. They suggest the Aryans fought against native peoples, whom they conquered and assimilated into the lower rungs of society. Later poets collected stories about these times into vast epics like the *Mahabharata* and the *Ramayana*, both of which recount the adventures of legendary heroes. Like the Vedas, these sagas were products of a martial nobility that lorded it over a peasantry that supported their lifestyle.[4]

The songs of the Vedas provide clues about how the Aryans saw their world. A panegyric to the sky god, Varuna, in the most ancient part of the Rig Veda, tells how he 'struck apart the Earth and spread it beneath the Sun as the priest, who performs the sacrifice, spreads out the victim's skin'.[5] In other words, the universe consisted of two halves, the Earth and the heavens, with the Sun occupying the atmosphere between them.[6] In later texts, like the *Bhagavad Gita*, the Earth, sky and atmosphere were referred to as the 'Three Worlds'.[7]

Other sections of the Vedas provide more details. The sky was said to bend over the Earth like a dome, stars sprinkled on its inner surface as the Sun and Moon floated within it.[8] Outside the universe, an inestimable void was shut out by the boundary of heaven and earth (together expressed as one Sanskrit word that means 'the whole world').[9] There is no mention of the five visible planets, although they did later become associated with gods mentioned in the Rig Veda.[10] Because it could not travel under the Earth, the Sun turned its face from the Earth at night as it returned to the place it would rise.[11] Solar eclipses occurred when a demon by the name of Svarbhanu enveloped the Sun in darkness.[12]

The constitution of the universe was not a central concern of the Vedas, but astronomy certainly was. The roster of sacrifices demanded by the texts had to be performed at the correct time, which meant the priests needed a method of accurate timekeeping. This task belonged to astronomers, who were an essential part of the team that ensured rituals were discharged by the book. The art of keeping time by observing the heavens was called *jyotish*, a profession recognized by the Vedas themselves. According to a guide from the fifth century BC, 'Sacrifices have been set out according to the sequences of time, therefore only he who knows astronomy, the science of time, understands the sacrifices.'[13] *Jyotish* was initially restricted to arithmetical formulas to determine the time of year and did not seek to describe the physical structure of the world.

Babylonian astronomers took a similar instrumentalist view of their art. Predictably, there has been heated debate among scholars about whether the earliest Indian astronomy owed anything to the techniques used in Babylon.[14] There were plenty of opportunities for Indian contact with Mesopotamia to the northwest. Traders had been visiting the powerful urban culture of the Indus Valley since the second millennium BC. Nonetheless, early guides to *jyotish* make no mention of the five planets, which contemporary Babylonians were well aware of. This militates against direct influence but, as we'll see, Indians didn't hesitate to appropriate foreign ideas they thought might be useful.

In the sixth century BC, the Persians invaded and held the Indus Valley as a satrapy until Alexander the Great took it over a couple of centuries later. His empire fell apart rapidly, but the Greek cities he had planted in Afghanistan kept Hellenistic culture alive in the east for several centuries thereafter. Modern Kandahar was among his foundations and was once called Alexandria Ariana. Meanwhile, the Maurya Empire acquired Alexander's possessions on the Indus in 303 BC, allegedly in exchange for five hundred war elephants. Under Emperor Ashoka (*c.* 304–232 BC), the Mauryans ruled almost the entire subcontinent. It took five hundred years for another kingdom, that of the Gupta dynasty, to finally reunite much of India, and then only briefly.[15]

The straightforward world picture of the Vedas, with heaven above and the earth below, was embellished in the Puranas. *Purana* is Sanskrit for 'ancient', so the term applies to a vast collection of old books, written mainly during the first millennium AD. They record myths and legends, genealogies and religious wisdom. Because the literature of the Puranas is so enormous and varied, it presents a multiplicity of world pictures. This was one of the things that the astronomer Lalla found disconcerting – one could take from the Puranas that the Earth sat on a turtle, a snake or a pachyderm. Some of the ideas about the shape of the world would become extremely influential in India and beyond. According to a core tradition in the Puranas, our world consisted of a circular continent surrounded by a saltwater ocean. In the centre of this landmass stood Mount Meru, the axis of the universe around which the heavenly bodies orbited. At dusk, darkness fell when the Sun disappeared behind the mountain, plunging the inhabitants to its south into shadow. It seems likely that Mount Meru and the Zoroastrian Mount Mara, both of which stood in the middle of the Earth, shared the same prehistoric Indo-European archetype.

Six other continents formed rings around the mountain, separated by oceans of various fluids such as butter or milk. The concentric bands of land and ocean stretched some 16 million kilometres (10 million mi.) across. The universe as a whole was a cosmic egg with fourteen horizontal layers within. The Earth's surface was the seventh storey, with six hells beneath it. Above, there were seven heavens, the lowest of which was the visible sky over our heads.[16] Later, we'll be meeting other examples of tiered world pictures from traditions across Asia.

The vast universe of the Puranas provided a stage upon which the many episodes of Indian mythology played out. Perhaps its size derived from the need to accommodate so many of the tales that formed part of the developing culture of Hinduism. For example, the fifth encircling ocean, containing milk, was the subject of a well-known myth in the Puranas, also recounted in the epic Mahabharata. The tale tells of how the gods wanted to churn the celestial sea of milk, perhaps the Milky Way, to precipitate the elixir of immortality. They used a mountain as a paddle, turned by an enormous serpent. When the elixir began to excrete, the gods formed up to receive their dose. The demon Svarbhanu, who featured in the Rig Veda as the cause of eclipses, disguised himself to join the queue but was spotted by the god Vishnu, who struck off his head. Unfortunately, the wily Svarbhanu had managed to get a sip of the elixir before this happened and so remained immortal, albeit in two pieces. Later, his head came to be called Rahu and

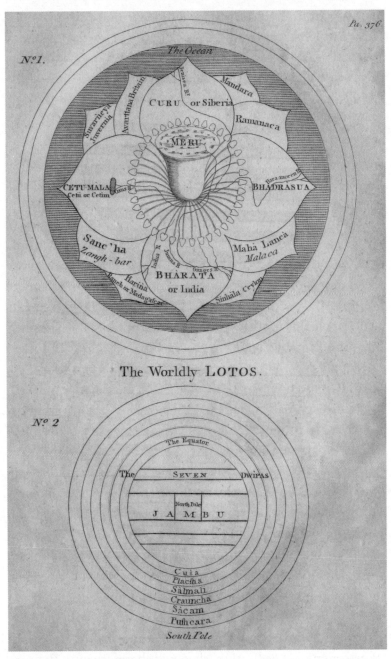

The world of the Puranas, from Captain F. Wilford, 'An Essay on the Sacred Isles in the West . . .', in *Asiatic Researches*, vol. VIII (1808).

his body Ketu, both of which were identified with planets.[17] Rahu and Ketu were dark and so couldn't be seen. When they obscured the Sun or Moon, they caused eclipses, much like their conjoined form, Svarbhanu, had done in the Rig Veda.[18]

The Globe Comes to India

The imperative to perform the Vedic sacrifices at the appointed times drove the need for greater astronomical precision. The Brahmin priesthood needed a reliable calendar, which could be computed from a proper understanding of the periods of the Sun and Moon.[19] At the same time, astrology, with its enticing promises to make sense of the world, grew more popular in India. Astrologers were more demanding than the Brahmins. They required techniques to track the movements of all the planets against the background of the stars. The pressure for accuracy from both groups drove progress in Indian astronomy.

Indian astronomers recognized that their status and ability to attract wealthy patrons depended on having the most innovative techniques. This requirement forced them to look to foreign expertise. Sometime after AD 1, knowledge of the planets and the twelve signs of the zodiac had arrived in India from Mesopotamia. The Sun, Moon and five visible planets combined with the native Indian tradition of the two dark planets Rahu and Ketu, giving nine heavenly bodies in all.

However, it was the wisdom of the Greeks that was the most prestigious. The *Mahabharata* described them as 'all-knowing' and they were especially admired for their knowledge of astronomy.[20] 'The Greeks are indeed impure,' a writer from the early first millennium AD commented, 'but among them this science [of astronomy] is well-established and therefore even they are honoured as accomplished people.'[21] This is quite a compliment.

By calling the Greeks 'impure', the author meant that they were unbelievers, but he still called them 'accomplished', a word often used to describe the authors of sacred books.

The earliest datable Indian treatise that sets out a world picture influenced by the Greeks was the work of an astronomer called Aryabhata (AD 476–c. 550). He lived in Pataliputra, the ancient capital of the Maurya emperor Ashoka, in northeastern India on the River Ganges. By Aryabhata's time, it was ruled by the Gupta dynasty.[22] His short treatise covers the calendar and mathematics before summarizing Greek-style spherical astronomy. Aryabhata was unambiguous about the shape of the Earth, echoing Aristotle when he said, 'The sphere of the Earth, being quite round, [is] situated in the centre of space, in the middle of the circle of the constellations, surrounded by the orbits of the planets, consists of water, earth, fire and air.'[23] He goes on to correctly explain eclipses as occurring when the Moon's disc occults the Sun, or the Earth's shadow passes across the Moon. He doesn't need to invoke the dark planet Rahu of popular lore.

In his book, Aryabhata clearly summarized the Aristotelian world picture. He arranged the planets according to the most common Greek sequence, with Mercury and Venus closest after the Moon, then the Sun, and finally Jupiter and Saturn. However, he combined his Greek material with elements from the Puranas. He located Mount Meru at the North Pole, the centre of the land: 'In the middle of the Nandana Forest is Mount Meru, five miles high and across, shining, surrounded by the Himavat Mountains, made of jewels, quite round.'[24] Hell was at the South Pole in the midst of a great ocean, while the mythical city of Lanka was on the equator. Mount Meru's demotion from the axle of the universe to a mere 'five miles high', or 8 kilometres, was quite precipitous, but it retained its symbolic station as one of the points through which the axis of the universe appeared to turn.

In some ways, Aryabhata's world picture may have been more modern than Aristotle's. In common with a few Greek thinkers, he suggested that the Earth itself rotates. 'As a man in a boat going forward sees a stationary object moving backward,' he wrote, 'so at Lanka [on the equator] a man sees the stationary constellations moving backwards in a straight line.' This implies that the man at Lanka is moving while the stars are not. But in the very next line, Aryabhata seemed to contradict himself: 'The cause of the [stars] rising and setting is the fact that the circle of constellations together with the planets, driven by the wind, constantly moves straight westward at Lanka.'[25] Whatever Aryabhata thought, other Indian astronomers were unimpressed by the idea that the Earth turns. They either ignored the idea or attacked the concept as nonsense. As Lalla, writing in the eighth century, complained, 'If the Earth rotates, how could birds come back to their nests? Moreover, arrows shot in the sky would fall towards the west.'[26]

Notwithstanding disputes about the Earth's motion, Indian astronomers accepted the basic system that Aryabhata had set out. At least half a dozen Sanskrit astronomical manuals adopted the Aristotelian world picture in the centuries after his death. That's not to say they agreed on the details. There is plenty of variation in the parameters that they use for their calendars and calculations of the planets' movements. Several schools emerged, each with a different set of observations and conclusions. Frustratingly, no records of these observations survive, although there are literary references to the astronomical instruments that were in use at the time.[27]

While it is certain that the theory of the Globe originally had a Greek origin, the route it took to India is unclear. Indian astronomers didn't have access to the work of Ptolemy with all his geometry and tables, but that wasn't a problem for them.[28] They were more than capable of doing the necessary astronomical observations and calculations themselves. Instead, they probably heard an outline of Greek cosmology, combined it with some traditional Indian beliefs and then used their own mathematical skill to fill in the gaps. There's no shortage of evidence for the penetration of Greek astronomical ideas into India around the time that Aryabhata was working. For example, one surviving Sanskrit text on astronomy is called the 'Treatise of the Romans', another is attributed to a certain 'Paul', who is likely to be a Greek.[29] There is also an extant book called the *Yavana-jakata*, which explains the art of astrology 'according to the instruction of the Greeks', although it has nothing to say about the shape of the Earth.[30]

It's suggestive that there had been thriving Greek kingdoms just to the north of India. As it happened, a Greek ambassador called Megasthenes (*c.* 350–*c.* 290 BC) had reached the Maurya empire shortly after Alexander's death and wrote about his experiences. His account doesn't survive but is quoted by later authors, including the geographer Strabo. According to Strabo, Megasthenes noted similarities between the world pictures of the Greeks and the Indian Brahmins. For example, both groups thought the cosmos was spherical, with Earth at its centre.[31] This could reflect the statements in the Puranas that the universe is egg-shaped, with the Earth in the middle of a stack of fourteen worlds. However, his testimony is not evidence of an awareness in India at this time that Earth is a sphere.

The Greek colonies planted by Alexander the Great grew into an empire that controlled parts of Afghanistan and northwest India until about AD 10. However, it had declined centuries before Aryabhata lived in the fifth century, so is too early to have been a significant influence on him. We need to search for later contacts to make an educated guess about how the Globe reached the subcontinent. A clue comes from coins discovered in India. Under the Gupta Empire, India grew rich by exporting luxuries such as silk and ivory to the Roman Empire. Much of this exchange was by ship, between the Malabar Coast and harbours on the Egyptian shore of the Red Sea. Strabo noted, in the first century AD, that 120 vessels left for India annually from just one port, whereas previously just a few had sailed. Evidence of this trade has been found right along the west coast of India.[32] Fishing villages expanded into ports, attracting craftsmen whose workshops produced wares for export. In return, the Romans shipped wine, oil and glassware but still suffered from an abysmal trade deficit with the Indians. They bridged the gap by purchasing Indian goods with gold coins. The Roman government made efforts to combat the outflow of bullion with quotas and tariffs, but the desire for eastern luxuries by the upper classes was insatiable and these mercantile efforts failed.[33]

Sailing through the treacherous waters of the Red Sea and across the Indian Ocean required the ability to navigate. A Greek guide to the eastern sea lanes survives, but a ship's captain would also have required enough knowledge of the stars, as well as the Sun and Moon, to help him find his way when the coast was out of sight.[34] He wouldn't have bothered with the work of mathematical astronomers such as Ptolemy or Hippocrates, but it is very possible that simpler textbooks, like those of Geminus or Cleomedes, reached India on a trading vessel. No Indian translations of these particular books survive, but the *Yavana-jakata*, the Sanskrit astrology manual noted above, looks like it is based quite closely on a Greek original.[35] Indian astronomers would have had little trouble rebuilding the mathematical superstructure for a geocentric universe and compared the results to their observations. We can't be certain exactly how this process happened, but it would explain why India came to adopt the Greek cosmos without also sharing Western mathematical parameters.

By the seventh century AD, Indian astronomers agreed the Earth is round. However, they recognized that the sacred cosmology of the Puranas had to be respected. We've seen how Aryabhata found a place for Mount Meru at the North Pole. Lalla, who was trenchant in his criticism of a flat earth and the shadow planets Rahu and Ketu, still did his best to incorporate as much from the Puranas as he could. He tucked the six lower layers of the universe, traditionally depicted as various kinds of hell, within the Earth and exiled other aspects of sacred geography safely away in the Southern Hemisphere.[36] The prestige of *jyotish*, conceded in the scriptures themselves, as well as the important job astronomers did when they ensured the accuracy of the calendar, helped propel acceptance of the Globe. At some point, the Greek world picture was even incorporated into one of the Puranas.[37]

Consensus among the astronomers didn't necessarily translate into widespread acceptance among the wider population, even for those who were literate. For most people, the scriptures remained authoritative and, by the fifteenth century, astronomers were on the defensive. They responded by being increasingly creative in their interpretations of the Puranas. For example, the *Vyanasiddhanta*, dating some time before AD 1500, combined the traditional view of a flat earth with astronomical theories about the sky and planets.[38] Jnanaraja (*fl. c.* 1500), the patriarch of a distinguished family of astronomers, took a different approach. He attacked those 'low-minded people' who rejected arguments for the Globe, and said those who used the Puranas as proof texts for a flat earth didn't understand them.[39] However, he accommodated his cosmology to the traditional view more closely than Lalla had been willing to do. For example, he affirmed that the Earth, while spherical, is supported by a tortoise, being an incarnation of the god Vishnu. He placed the seven encircling oceans in the Southern Hemisphere, except the familiar saltwater one, which is in the north. Finally, he rejected gravity as the reason objects don't fall off the Earth, preferring a combination of divine intervention and the special ability that antipodeans possessed to help them adhere to its surface.[40]

In the eighteenth century, writers became ever more resourceful in their efforts to mix tradition with astronomy. One author postulated that the huge circular world with Mount Meru at its centre, ringed by continents and oceans, was real. A much smaller spherical earth floated above, with the planets orbiting it. Others decided to reassert the traditional world picture and try to work astronomy into it as best they could. A further suggestion was that the world in ancient times had indeed been flat, but an ancient cataclysm gave us the spherical world that we see today.[41] Luckily, the vast majority of working astronomers used tables and calculating aids that didn't require them to worry about cosmology. All that mattered to them was being able to accurately predict the motion of the lights that we see in the sky. With tables keyed to their location, they could determine the calendar, the timing of important Hindu rituals and even produce horoscopes using purely arithmetical methods.[42]

India's central position in South Asia meant that its culture spread widely. In particular, it had close trade connections with the Persian Empire to its northwest which, by the second century AD, was under the control of the Sassanid dynasty. The Persians had heard about the Globe, but the tiered world picture of the Puranas would be a more profound influence on them.

12

The Sassanian Persians: 'Good thoughts, good words, good deeds'

Once upon a time, an envoy from India, leading a convoy of elephants laden with treasure, arrived at the court of the Persian Shah of Shahs, Khosrow I (*c.* AD 512–579). The Indian carried with him a chess set with pieces made from ruby and emerald. He declared that he would hand over all his riches if there was one among the Persians who could beat him at the game. The challenge was made harder because the visitor declined to explain the rules. For three days, the councillors of the shah examined the chess set and tried to understand the logic of the game. Just as they were going to admit defeat, a certain Wuzurgmihr, son of Boxtag, came forward claiming he would reveal the secret of chess. As good as his word, he described the moves made by each piece and then beat the Indian three times. Khosrow was delighted. Wuzurgmihr then introduced his own game, which he called backgammon, for the embassy to take back to India. He explained that the backgammon board was level like the Earth, while the black-and-white pieces stood for night and day. The dice, with six sides, turned like the heavens around the Earth, with each number representing part of the creed of Zoroastrianism, the official faith of the Persian Empire. 'One' for the supreme God; 'two' for the spiritual and material worlds; 'three' for 'good thoughts, good words, good deeds'; 'four' for the four quarters of the Earth; 'five' for the various lights in the sky; and 'six' for the periods of creation. Khosrow clapped his hands in pleasure and ordered Wuzurgmihr to accompany the envoy back to India carrying gifts to demonstrate the superior wisdom of the Persians.[1]

The Sassanian Empire

This tale is preserved in Middle Persian, the language of a second Persian Empire that stretched from Afghanistan to the River Euphrates. It was ruled by the Sassanid dynasty and arose five centuries after Alexander the Great had conquered the first one. The first Sassanid Shah of Shahs was Ardashir I (AD 180–242), who defeated the Parthian king Artabanus IV (AD 191–224) in battle and took over his realm.

Under the Sassanids, the Zoroastrian priesthood expected that all Persians should follow their rituals and teachings. In particular, they emphasized the duty of the shahs to follow the faith as they laid it down. 'Kingship is arranged according to religion and religion is based on kingship,' the priests declared, binding themselves tightly to the state.[2] But enforcing doctrinal orthodoxy was a challenge. Although the priests had carefully collated the ancient words of the Avesta, these oral traditions were as much as 2,000 years old by then. Long regarded as sacred, their form had been frozen in a forgotten language. Consequently, almost no one could understand what they meant.[3] Modern scholars can make a better fist of it by comparing the Avesta to other early Indo-European texts like the Rig Veda, but much is still obscure. The Sassanian priests attempted to promote their interpretation through commentaries, but that didn't stop schisms wracking Zoroastrianism over the next few centuries.

In addition to Zoroastrians, the Persians found themselves ruling substantial populations of Christians, Jews, Buddhists and other communities. Within this maelstrom of creeds and dogmas, brand new religions could arise from a compound of the doctrines of the old. The most successful example of this syncretism was a sect founded by a man called Mani (AD 216–277), born and brought up near Babylon. He co-opted Jesus of Nazareth as a prophet and drew on elements of Jewish and Vedic thought, the latter picked up during his travels to India. Manichaeism stressed the dualistic nature of Zoroastrian doctrine and declared that the material world was intrinsically evil. Its adherents sought to transcend the corruption of their physical bodies and ascend to a purely spiritual existence.

While the priesthood expected the shahs to be faithful disciples of the Avesta, rulers had larger concerns than clerics. The vast Sassanian Empire embraced many cultures and traditions, so trying to enforce religious uniformity risked sparking revolts. In any case, Shapur I (AD 215–270), Ardashir I's son and the second Sassanid shah, was sympathetic towards Mani and his followers. To the disgust of the priesthood, the shah allowed Manichaeism to spread across Persia and then into the Roman Empire. Unfortunately for Mani himself, Shapur's successors were less well disposed to his ideas and, perhaps wary of his growing influence, had him killed. Manichaeism survived his death, despite sporadic persecution across the Persian and Roman Empires. In the West, its most famous votary was St Augustine of Hippo (354–430), a Christian bishop and saint, whom we will meet later. To the east, it spread through Central Asia and into China, where it survived until at least the fourteenth century.

The chance survival of a remarkable painted scroll, created in China in about AD 1300, allows us to admire the Manichaean universe in all its glory and complexity. The scroll, now housed in a private collection in Japan, was only identified as Manichaean in 2009. There are good reasons why it took so long to reveal its true origins. Its iconography is Chinese and it contains clear Buddhist elements. However, recent discoveries of Manichaean texts in the deserts of Central Asia have allowed scholars to recognize the correspondences between the painted scroll and these written sources.[4]

Finally seen for what it is, the scroll provides a singular window into the world of the Manichaeans, as well as their Zoroastrian and Buddhist antecedents. Originally, it would have been intended to hang on a wall as a teaching aid. It portrays a *Diagram of the Universe* in vertical orientation, about 1.2 metres (4 ft) tall. At the top is the 'Realm of Light' and just below it, the 'New Paradise', where blessed souls dwell. Next down are the ten celestial firmaments, with the belt of the zodiac in the lowest. Under the atmosphere are the eight tiers of the Earth. The highest is the surface on which humans live. The Earth is circular, divided into four quarters, and surrounded by a moat-like ocean. Beyond the Ocean is a ring of mountains, in which are nestled the fiery jaws of hell. The universe is held fast by pillars running up its side. In the middle of the Earth is a mountain range with what looks like an enormous tree at its centre. This is best identified as Mount Meru or Mara, the axis of the world in the Vedic and Zoroastrian traditions, with 32 cities occupying a plateau on top. An alternative interpretation is that the universe as a whole symbolizes a man, and this stiff upright object represents his phallus.

The shape of the world mattered to Zoroastrians and Manichaeans because it was the stage on which darkness and light fought their great war. Mani took for granted that a benign deity should dwell above us, while his adversary would lurk below. For Zoroastrians, the whole world had been created by the benign god, Ahura Mazda, but now it bore the scars of the struggle between good and evil. Manichaeans, however, incorporated the Earth within the realm of shadow, from which blessed human souls needed to escape. Both creeds, in common with Buddhist and Vedic traditions, imaged the universe as a multi-layered sandwich, with tiers of heavens above and hells below, and the Earth the filling in between.

It was essential to Manichaeism that there should be a hierarchy to existence, from high to low and from light to darkness, with the Earth caught up with base creation. However, it would have been quite possible for a spherical universe to express the same theological truths that are encapsulated in the *Diagram of the Universe*. Greek Neoplatonists had already successfully combined a hierarchical chain of being with an Aristotelian world picture. And, coincidentally, some of the leading lights of Neoplatonism visited Sassanian Persia in the sixth century A D.

The Closure of the Schools and Persian Exile

As we've seen, Neoplatonism began with the teaching of Plotinus in the third century AD. It was an eclectic philosophy that later drew some of its ideas from Christian thought. Christians returned the favour by incorporating elements of Neoplatonism into their theology. Nonetheless, many Neoplatonists were vehemently opposed to Christianity. The Roman Empire officially converted to Christianity in the fourth century and grew increasingly inhospitable to pagans thereafter. In 529, Emperor Justinian (482–565) promulgated a decree that the philosophical school of Athens, dominated by pagans, should close or at least lose its government funding.[5] The school called itself the Academy, in honour of Plato's original foundation, which had been closed down centuries earlier by Sulla. Justinian left the less obstreperous pagans of Alexandria alone, and philosophical teaching thrived there until Egypt was lost to the Roman Empire in the seventh century.

Newly unemployed, a group of seven Athenian teachers decided to take a chance on Persia. They had heard good things about the Shah, Khosrow I. As we saw in the tale about the origins of chess and backgammon, Khosrow was hungry for wisdom from India, although he probably sought knowledge of astrology rather than board games. On ascending the throne, the new shah had taken the precaution of wiping out much of his family to secure his position. More positively, his patronage of scholarship and translation, intended to shore up orthodox Zoroastrianism and strengthen the intellectual resources of his empire, earned him the moniker 'the immortal soul'. In particular, he was on the lookout for Greek texts to translate into his native language.

As far as the Persians were concerned, they were the ones who had first developed science and philosophy. Their legends held that, when Alexander the Great invaded in the fourth century BC, he had ordered that the books of Zoroastrian wisdom be translated into Greek. Then, he had the originals incinerated.[6] There is some truth to this. Alexander really did burn down the ancient Persian capital of Persepolis, resulting in incalculable cultural destruction. The Sassanid shahs had sponsored the collection of the surviving fragments of Zoroastrian scripture when they first came to power. Khosrow now wanted to regather the knowledge that Alexander had allegedly stolen and sent back to Greece.

As usual, it was books on astrology that the Persians were most interested in acquiring. By this time, the reputation of the Greeks as prognosticators had outstripped that of the original astrologers from Babylon. This was ironic because, as far as the Greeks were concerned, astrology was indelibly associated with the east. For instance, the magi said to have followed the star to the birthplace of Jesus in the Gospel of Matthew were astrologers and most likely Zoroastrian priests. The wide variety of Persian words that could describe an astrologer, including 'star teller', 'zodiac teller' and 'star reckoner', show how important the subject was for them.[7] Of particular interest were case studies that featured the horoscopes of individuals accompanied by a description of their lives. This allowed correlations to be detected between an individual's fate and the stars they were born under. The Sassanians also translated Ptolemy's astrological work.[8]

The leader of the merry band of Neoplatonist exiles was a certain Damascius (*fl.* 500–540), who'd previously been the head of the Athenian school, and among their number was Simplicius of Cyrene (*fl.* 520–60). Simplicius was the author of commentaries on the works of Aristotle that preserve valuable excerpts from lost books by the Presocratics. We were using these Presocratic fragments when we considered the work of Anaxagoras and Parmenides.

We know about the emigration of the seven Neoplatonists thanks to an account by a Greek historian called Agathius (*c.* 536–c. 582), written fifty years after the event.[9] It cannot have been an easy journey for the philosophers, since the Persian and Roman Empires were either at war or observing an uneasy truce. Still, even though Khosrow had staff at his disposal to translate texts from Syriac and Greek, the prospect of a visit from some of the most celebrated philosophers of Athens must have been enticing. They were welcome to settle in his empire.

When the Athenians arrived at the Sassanian court in the early 530s, the young monarch was already a seasoned military campaigner, an astute and ruthless politician and a noted scholar. It's likely he asked them questions in an effort to gauge whether they brought any useful scholarship from Greece. In fact, a treatise called *Answers to Khosrow* survives under the name of Priscian of Lydia (*f. c.* 500–c. 540), named by the historian Agathius as one of the seven philosophers. It is a series of frequently asked questions on Neoplatonic ethics and natural philosophy, and assumes the Aristotelian world picture with a spherical earth at its centre.[10] It's hard to determine whether Priscian presented the *Answers* to the shah himself, but the content is likely to be genuine.

The *Answers to Khosrow* shows Priscian of Lydia brought the Globe to Persia, but he was certainly not the only Greek to have done so. There were already plenty of others living in the Sassanian Empire. Descendants of Alexander's conquerors still dwelt in settlements stretching all the way to India. More recently, Shapur I had founded the city of Gundashapur in southern Iraq and populated it with the Greeks he had deported from Antioch after sacking the city in 253.[11]

Nonetheless, commentaries on the Avesta and more popular Persian literature, like the story of backgammon, show no awareness of the Globe. Even though the majority of surviving Zoroastrian texts date from the eighth or ninth centuries, long after the fall of the Sassanids to the Arabs, they preserve a traditional world picture. By the time these books were written, Zoroastrianism was a marginalized faith without any official influence. It is possible that the texts are deliberately anachronistic, intended to preserve ancient teaching that was in danger of being lost forever. In any case, there is no indication that the Globe ever became part of Zoroastrian thought, even though the tables used by Persian astrologers utilized spherical astronomy from India and Greece.[12]

As for the seven Neoplatonists, they didn't stay long in Persia. According to Agathius, they were appalled by their hosts' promiscuity and viciousness. We shouldn't rely too much on Agathius as an objective witness. He inaccurately described Khosrow himself as an ignorant buffoon. But the Neoplatonists were suffering from culture shock that made them long for home. One day, they came across a body lying unburied on the ground.

It had likely been left there in observance of the Zoroastrian rite, under which the dead were exposed to the elements until their bones had been picked clean by wild animals. The philosophers, behaving like a bunch of insensitive tourists, ordered their servants to bury the body. When they passed the same spot the next day, they found it had been exhumed again, doubtless by the dead man's family, appalled at the sacrilege the deceased had suffered. Luckily, one of the philosophers dreamed the corpse belonged to a man guilty of sleeping with his mother and the Earth itself had refused to cover him. So, they left it be and hurried home. Any amount of persecution by Justinian was preferable to the incomprehensible rituals of the Persians.[13]

Meanwhile, the Sassanian Empire continued its struggle for supremacy with the Romans. Under Khosrow II (c. 570–628), grandson of Khosrow I, the two empires fought a titanic war in the course of which the Persians occupied Syria and Egypt. The struggle left them bankrupt and exhausted, unable to resist when armies fired with zeal for their new creed of Islam swept out of the Arabian Desert in the mid-seventh century. Today, only a few communities of Zoroastrians sing the hymns from the Avesta, but tales of the Sassanid dynasty are still remembered in the *Book of Kings*, Iran's national epic.

13

Early Judaism: 'From the ends of the Earth'

The Roman emperor Hadrian (AD 76–138) certainly got around. In his twenty-year reign he visited Britain, Egypt and most of the places in between. According to Jewish tradition, while in Judea, he met a wise rabbi called Joshua ben Hananiah (d. AD 131) and bombarded him with questions. One topic of conversation was the gestation period of snakes. Joshua ben Hananiah said it was seven years, but Hadrian insisted that the pagan philosophers of Athens had declared it was just four years. To resolve the dispute, he ordered the rabbi to travel to Athens and bring the philosophers back to the emperor's presence.

Joshua ben Hananiah sailed to Greece to find the pagans' headquarters. After he had tricked the guards, he entered their lair and asked to learn of their wisdom. They responded by challenging him to a riddle competition. If the rabbi won, the philosophers promised to come and have lunch with him on his ship. Otherwise, they would kill him. The contest started and riddles fizzed back and forth between the Jew and the pagans.

'If the chick inside an egg dies,' the philosophers asked, 'which way does its spirit emerge?'

'It goes out the way it came in,' the rabbi responded.

'If salt goes bad,' demanded the philosophers, 'what do they salt it with?'

'With the placenta of a mule,' said the rabbi.

'But a mule [being sterile] doesn't have a placenta,' said the philosophers. Joshua ben Hananiah was unperturbed. 'Does salt go bad?' he retorted.

'Where is the centre of the world?' they asked.

'Right here!' replied the rabbi. Perhaps this is a reference to a particularly odd tenet of Greek philosophy: that the world is a sphere and so its surface has no centre.

Eventually, the philosophers admitted defeat and agreed to have a meal with Joshua ben Hananiah on his boat. But as soon as they were aboard, he locked them up and shipped them back to Hadrian. On arrival, they were impertinent to the emperor, who asked the rabbi to do away with them.[1]

The World Picture of the Bible

Joshua ben Hananiah probably did meet Hadrian. Pretty much anyone who was anyone had a chance to say hello to him during his tours of his empire. Unfortunately, the good relations between the emperor and the rabbi were of no help to the Jews. In AD 131, shortly after the rabbi's death, Judea revolted, allegedly over Hadrian's attempt to ban circumcision. The Romans crushed the uprising, exiled the population and renamed the province Palestine.

While there is some history behind it, the tale of Joshua ben Hananiah and the philosophers is a tall story intended to demonstrate that pagan Greek sages weren't nearly as clever as they thought they were, certainly compared to the crème of the rabbis. The story appears in the Talmud, which was first written down around AD 600. It contains a profusion of sayings by learned Jews, expounding and clarifying matters of religious law. The standard printed edition weighs in at 6,200 pages, making it the largest work to survive from antiquity.[2] Furthermore, there are two versions of the Talmud, one prepared in Palestine and the other in Babylonia. There was a Jewish community in

Babylon because the population of Jerusalem had been deported there by King Nebuchadnezzar in the sixth century BC. This exile ended in 539 BC, when Babylon in turn fell to the Persian Cyrus the Great. He allowed the Jews to return to Jerusalem, although many elected to stay in Mesopotamia, where they had grown prosperous and influential.

Jews venerate the *Tanakh*, or Hebrew Bible, called the Old Testament by Christians. This collection of books had reached its final form by 300 BC but drew on traditions that went back a thousand years earlier. Given its antiquity, it is to be expected that the Hebrew Bible reflects a traditional world picture. That the Earth is flat isn't something that the scriptures explicitly teach; it is simply an assumption that the biblical authors held in common with everyone else at the time.[3] For example, we've seen how, in the eighth century BC, the Hebrew prophet Isaiah had referred to the 'four corners of the world', probably to mock the ambitions of the Assyrian monarchs who were menacing Jerusalem. References to the 'ends of the Earth' are more common in the Bible than to its corners. For example, Job proclaimed God 'looks to the ends of the Earth, and sees everything under the heavens' and 'stretches out the heavens' over the world (Job 28:24 and Job 9:8). Likewise, Job noted that the Earth is supported by pillars, which God shakes to cause quakes (Job 9:6). As Isaiah said, 'It is [God] who sits above the circle of the Earth, and its inhabitants are like grasshoppers; who stretches out the heavens like a curtain, and spreads them like a tent to dwell in' (Isaiah 40:21–2). Isaiah's reference to 'the circle of the Earth' can only mean a disc, not a sphere, in this context.

The account of the six days of creation in the Book of Genesis was one of the last parts of the Bible to be composed. The author, usually assumed to be a Hebrew priest, subscribed to a world picture similar to the Babylonian one. The parallels with how Marduk created the world in *Enuma Elish* are not exact, but it is clear that both accounts arose from a common Mesopotamian milieu. However, the Babylonian and Hebrew authors had different theological agendas. *Enuma Elish*, as we have seen, sought to justify the domination of Babylon over the world. By contrast, the Book of Genesis portrays creation as an act of benevolence by God, who fashioned humanity in his own image.[4]

Given the provenance of the Bible, it can't be expected to include anything about the Greek doctrine of the Globe. After all, Aristotle only set out the evidence that the Earth is a sphere in the mid-fourth century BC, by which time most books of the Hebrew Bible had already been written. It would be as anachronistic to say it teaches the Earth is a ball as it would be for it to state the Earth orbits the Sun. Besides, the biblical authors weren't interested in educating anyone about the constitution of the universe. They wrote in everyday language that was easy to understand, so they could communicate the core doctrine that God ruled the heavens and the Earth.

Other ancient Hebrew writings, which didn't make the cut when the Bible was collated, were more explicit about the shape of the world. For example, according to the Book of Enoch, composed around 300 BC, the Sun rose from one of six gates at the eastern edge of the Earth, crossed the sky and set through a portal in the west. At night, the Sun returned to the gates of dawn by travelling behind the firmament around the northern rim of the Earth. The gates through which the Sun rose and set varied through the seasons, which accounted for its different paths across the sky.[5] Although the Book of Enoch wasn't included in the Bible, it was a respected text and has been found among the Dead Sea Scrolls.[6] Other Jewish writings, with names like the Apocalypse of Ezra and the Testament of Abraham, contain alternative world pictures derived from Babylonian lore, but none mention the Globe.[7]

The Talmud

By the time the rabbis quoted in the Talmud were ruminating over sacred law in the early centuries AD, knowledge of Greek philosophy was widespread around the Mediterranean basin. However, Jews had good reason to be wary about engaging with it. In the second century BC, King Antiochus IV of Syria (c. 215–164 BC) had tried to impose Greek customs, including the worship of pagan gods, on the Jews in his realm. He was the ruler of the part of Alexander the Great's old empire that encompassed Judea and Galilee. His persecution set off a bloody revolt, during which revolutionaries successfully set up an independent polity ruled by the High Priest in Jerusalem. This meant that Jews in Judea and Mesopotamia had little love for Greek culture and many held it in contempt, as the story of Joshua ben Hananiah and the Athenian philosophers so vividly illustrates.

As we'd expect, the rabbis adopted a traditional world picture to elucidate the meaning of the Bible, taking little notice of Aristotelian cosmology. For example, those quoted in the Talmud concurred that the firmament mentioned in Genesis formed the roof of the sky. This raised questions about what the firmament is made of. One of the charms of the Talmud is that the rabbis it cites often disagreed, leaving the reader at a loss about the correct interpretation. When it came to the constitution of the firmament, some thought it was solidified water that had congealed, much like when milk curdles into cheese.[8] Others said it was fire and water whipped together. The number of firmaments was also up for debate. Despite lacking any biblical precedent, one influential passage of the Talmud said there were no less than seven of them, bounding seven heavens, each with their own names and functions, piled on top of each other.[9] We've already encountered seven heavens in the Vedic Puranas, but we shouldn't assume a link between India and the Talmud. It is far more likely that the concept derived from the seven visible planets – the Moon, the Sun, Mercury, Venus, Mars, Jupiter and Saturn.

While there's a consensus in the Talmud that the Earth is flat, there is lots of wrangling about the details. Some rabbis said its foundations were extremely deep – a distance that would take fifty years to travel. Alternatively, the Earth might rest on water, which in turn flowed over mountains. And the mountains stood upon the winds.[10] This meant the Sun couldn't pass under the Earth at night. Instead, as the Book of Enoch had stated, it travelled around the edge of world, behind the firmament.

On this last point, the Talmud quoted Judah ha-Nasi (c. AD 135–220), who, exceptionally, was curious about what the Greeks had to say respecting the matter. He noted gentile experts said the Sun passed beneath the Earth, and he thought they might be right about this. As evidence for his contention, Judah ha-Nasi observed that springs were warmer at night. Perhaps, he mused, this was because the Sun heated underground watercourses as it passed beneath them.[11] Judah ha-Nasi was such a revered commentator that he is often simply referred to as 'Rabbi' in the Talmud. We'll soon see how his concession that the Greeks were correct about the route of the Sun helped to reconcile later Jewish scholars to the Aristotelian universe.

Judaism and Greek Culture

The vicissitudes of history had bestrewn the Jewish people from Spain to India. One of the largest populations of the diaspora settled in Alexandria. There had been a Jewish community there since shortly after Alexander the Great founded the city, and by the first century AD it had become well established. It was natural that they spoke Greek rather than the Aramaic of the Middle East, let alone the Hebrew of the scriptures. As a result, they needed a copy of the Bible in their own language.

The Greek translation of the Tanakh, completed before 100 BC, is called the Septuagint, meaning 'the seventy'. It was authoritative for Greek-speaking Jews and, later on, for Christians as well. The name derives from the legend about its creation. According to the story, King Ptolemy ordered the head librarian of the Great Library of Alexandria to create a comprehensive collection of knowledge. 'The president of the king's library', the earliest source of the story explained, 'received vast sums of money for the purpose of collecting together, as far as he possibly could, all the books in the world. By means of purchase and transcription, he carried out, to the best of his ability, the purpose of the king.'[12] As part of this project, the librarian arranged for a crack team of 72 Hebrew scholars to be brought from Jerusalem to translate the Bible.

Even though it was written in Greek, the Septuagint no more acknowledged Aristotelian cosmology than the Tanakh had done. In fact, where it diverges from the Hebrew Bible, it cleaves more closely to the world picture in the Book of Enoch. For instance, in a passage in the Book of Ecclesiastes, the original Hebrew reads, 'The Sun also arises and the Sun goes down, and hastens to its place where it rose.' The translators of the Septuagint rendered this passage, 'And the Sun rises, and the Sun goes down and draws toward its place; arising there it proceeds southward, and goes round toward the north' (Ecclesiastes 1:5–6). This seems to mean that during the day, the Sun travels in the southern sky, while at night it skirts back to where it rises around the north edge of the world. Greek-speaking Christians would later use this infelicitous translation of Ecclesiastes as evidence that the Earth is flat.

The diverse population of Alexandria made the city a fractious, violent place. In AD 40, things had got so bad that a delegation of Alexandrian Jews travelled to meet the emperor in Rome to complain about the havoc wrought by the local pagans. It was led by a respected scholar called Philo (*fl. c.* AD 20–c. 40). His younger brother, a friend of kings and emperors, had become almost completely Hellenized, but Philo continued to straddle Jewish and Greek culture. He had benefited from a philosophical education that included readings in the Stoics and Aristotle but saved his passion for Plato. Much of his surviving work attempts to synthesize Plato's philosophy with the Bible by imputing figurative meanings to the sacred text. For example, Genesis tells how God set angels to guard the Garden of Eden after he had expelled Adam and Eve. Philo allegorized the angels with reference to Greek astronomy. He suggested one of them was an allusion to the outer sphere of the cosmos turning from east to west, and the other to the inner planetary spheres that rotate from west to east.[13] It's clear Philo subscribed to the Aristotelian world picture: the seven planets orbit the Earth within a spherical universe. He accepted that Earth is a ball, referring in passing to one of its hemispheres, and that the Sun passes under it at night.[14] But he did not analyse or seek to refute the passages in the Bible that imply the Earth is flat. He was content to leave the issue undiscussed.

Philo's acceptance of Greek natural philosophy, Globe and all, was not the norm among Jewish people of his time. One example of a Greek-speaking Jew who stuck to the traditional world picture was Flavius Josephus (AD 38–100), best known for his chronicle of the disastrous rebellion against the Romans that culminated in the destruction of the Temple in Jerusalem in AD 70. After this cataclysm, Josephus moved to Rome, where he composed books that tried to defend Judaism from its pagan denigrators. His history of the Jews, based on the Bible, argued that they were an ancient race as worthy of respect as Babylonians, Egyptians and Romans. Josephus wrote in Greek for the benefit of his audience and drew on exemplars from literature rather than philosophy. His world picture resembled Homer's description of a disc surrounded by ocean, so he had little trouble reconciling it to the Bible. For example, his retelling of the creation story in Genesis described how God poured the sea around the Earth so that it is encircled by water, much like in Homer's passage on the Shield of Achilles in the *Iliad*.[15]

In the following centuries, Greek-speaking Jewish communities gradually assimilated into the Judaism of the rabbis, who privileged their own writings in Hebrew and Aramaic. The influence of Greek literature was marginalized and Philo's syncretism of Judaism and pagan philosophy forgotten. The rabbis continued to oppose Greek cosmology and to teach a traditional world picture. Nonetheless, they sensed that this was a subject that had to be treated with caution. The law forbade lectures on the beginning of Genesis, which was to be taught to students individually. The creation story contained secret meanings that the rabbis believed should only be imparted to those trained to receive them.[16] But this situation didn't last forever. Centuries later, when Islam had mastered the Middle East, Jewish theologians would become acquainted with Greek philosophy, as well as knowledge of the Globe, from Arabic scholarship.

We'll come to that story soon, but first we need to see how the Globe was treated by a group of Jewish sectarians who identified a Galilean preacher called Jesus of Nazareth as their founder. These were the Christians.[17]

14

Christianity: 'All things established by divine command'

On 7 April in the year AD 30, the Roman governor of Judea, Pontius Pilate (*fl.* AD 26–36), ordered the crucifixion of a Jewish troublemaker outside Jerusalem.[1] There was nothing very unusual about this. Rome had subjugated the Mediterranean basin and now ruled it with an iron fist. Rebels could expect only agonizing death. Jesus of Nazareth (*c.* 4 BC–AD 30) had only been a mild irritant to the Romans by threatening to cause a disturbance in Jerusalem at the time of the festival of Passover. But that was reason enough for them to scourge him to within an inch of his life and then nail him to a cross to die.

Later, the followers of Jesus became convinced that he had been resurrected from the dead. They began to preach to their fellow Jews that he was the long-awaited Messiah. Regrettably, they said, a colossal misreading of the Bible had led the Jews to expect the Messiah to be a political leader, whereas, in fact, his kingdom was spiritual. Jesus' disciples were all Jews, but, following a debate, they decided to evangelize pagans as well. From a slow start, this initiative was tremendously successful and in the fourth century AD the Roman Empire itself became officially Christian.

Jesus and the New Testament

The New Testament, containing the four gospels together with various letters, was written, almost exclusively, by Greek-speaking Jews. As a result, it is filled with citations from the Septuagint, the Greek translation of the Old Testament made in Alexandria, and reflects its traditional world picture. An example is the story about the temptations of Jesus. According to the Gospel of Matthew, Jesus had been fasting in the wilderness for forty days before the Devil tried to corrupt him. After two failed attempts to bend Jesus to his will, he had another go. 'Again, the Devil took him to a very high mountain, and showed him all the kingdoms of the world and the glory of them; and he said to him, "All these I will give you, if you will fall down and worship me"' (Matthew 4:8). Jesus was having none of it, and the Devil relented.

The gospel's author assumed that you can see every land from a sufficiently lofty peak. This only makes sense if the Earth is flat, which is precisely what we would expect from a Jewish author writing in the first century AD. Likewise, the New Testament's Letter of St Jude, ostensibly written by one of Jesus' brothers, quotes from the Book of Enoch as if it was inspired prophecy (Jude 1:14). Presumably, the author of the letter would have taken Enoch's world picture at face value as well.

However, one of the evangelists was certainly not Jewish. This was the author of the Gospel of Luke and the Acts of the Apostles. He's traditionally identified as a gentile doctor who travelled with St Paul of Tarsus (d. c. AD 65) on some of his missionary journeys. There is very little reason to doubt this, not least because parts of the Acts of the Apostles are narrated in the first person, showing the author was an eyewitness. Luke was familiar with Greek literature and the various philosophical schools of his time. In the Acts of the Apostles, he told of

Paul's encounter with a group of Stoic and Epicurean philosophers in Athens, and quoted from the *Phenomena* of Aratus, the
popular poetical description of the constellations (Acts 17:28).

The Gospel of Luke habitually echoes stories in the
Gospel of Matthew, showing they share the same sources.
But it is where they differ subtly that's most revealing. For
instance, Luke recounted the story about the temptations of
Jesus as follows:

> And the Devil took him up, and showed him all
> the kingdoms of the inhabited world in a moment
> of time, and said to him, 'To you I will give all this
> authority and their glory; for it has been delivered
> to me, and I give it to whom I will. If you, then,
> will worship me, it shall all be yours.'
> (Luke 4:5–7)

Luke consciously changed the passage from the one found in
Matthew. He removed the mountain and replaced it with a
vision of the world laid out beneath Jesus' feet. And there is
another difference: the Greek word that Luke used for 'world'
is *ecumene*, which literally means 'the known world' of the three
continents of Europe, Asia and Africa. In his mind's eye, Luke
imagined the Devil displaying the inhabited parts of the Earth to
Jesus from space, just like in Cicero's *Dream of Scipio*. However,
when Matthew spoke of the world in his version of the temptation, he used the word *kosmos*, meaning the whole universe.
The difference in the language used by the two gospel authors
is best explained by Matthew following the world picture of the
Old Testament, while Luke knew the Earth was a sphere. Later
Christians would face a choice of which road to follow: embrace
the traditional view of the Bible, substitute it for Aristotle's Globe
or wish the entire problem would go away.

Early Christianity

Ancient writers cared little about what commoners thought about anything, so it is difficult to ascertain their opinion of a marginal topic like the shape of the Earth. Christianity initially found its converts among the lower classes who, if not completely illiterate, were extremely unlikely to be cognisant about cosmology. Their religious concerns were domiciliary as well. Before becoming Christians, they worshipped their household gods and honoured their ancestors.[2] While they might acknowledge the existence of the great deities like Zeus and Athena, they would hardly dare pray to them. That was the prerogative of high priests and emperors, not hoi polloi. Christianity's major selling point was that the Big God, the one who created the world and toppled empires, genuinely cared about ordinary folk. Christians claimed that the moral order of the universe didn't just benefit aristocratic warriors or wealthy intellectuals, but everyone else as well.

The best indication of what ordinary Christians thought comes from the various apocryphal books that didn't make it into the New Testament, with titles like the Apocalypse of Paul. They sometimes survive in languages such as Coptic or Syriac rather than Greek. Several feature tours of the cosmos, and from them we can glean that their anonymous authors assumed the Earth was a flat disc, perhaps surrounded by water, with the sky arched overhead.[3] Christians also read non-canonical Jewish writings like the Book of Enoch, with its gates at the edge of the world through which the Sun rose and set. These traditional world pictures persisted for centuries among ordinary people in the eastern Mediterranean, whether they were Christians or not.

Still, almost all surviving early Christian writing comes from the pens of bishops. Unlike most of their flock, these men were literate and often had a classical education. They were familiar with the writings of philosophers like Plato and Aristotle, even if at second hand, and had imbibed the Aristotelian world picture while at school.

The bishops understood the Bible implied a traditional cosmology and contained no reference to the Globe. This was part of a wider problem. How should they integrate the useful aspects of pagan knowledge into Christianity, while avoiding its dangers? There wasn't a consensus on how to handle this. Early Christian attitudes to Greek philosophy ranged from qualified enthusiasm to unmitigated condemnation.[4]

While most of the bishops knew of Aristotelian cosmology, they were mindful it could be a sore point among some of their flock. They needed to tread carefully to humour conventional opinions without conceding any credibility to them. St Basil (329–379), Bishop of Caesarea, dealt with the problem in a typical way. He was born to a wealthy family, educated in Constantinople and Athens by the leading pagan scholars of the day, and knew perfectly well how Aristotle and Ptolemy saw the cosmos. However, he was acutely aware that, when he ministered to plain folk, he should avoid confusing them with irreconcilable accounts about nature.

Sometime during the 370s, Basil delivered a set of sermons on the creation story in Genesis to a congregation of working people. In these homilies, he outlined the Aristotelian model of the universe and made it quite plain that he thought it was accurate enough. But that was beside the point. As he told his audience, 'If there is anything in this system which might appear probable to you, keep your admiration for the source of such perfect order: the wisdom of God.'[5] Similarly, when he considered whether the Moon's light is reflected from the Sun, he managed to dismiss the question while still making it clear he accepted the answer given by the philosophers.[6]

For centuries, educated Christians embraced this supercilious attitude towards their parishioners. Four hundred years after Basil of Caesarea had delivered his sermons on Genesis, St John Damascene (675–749) used a similar strategy in his *Exposition on the Orthodox Faith*. John was a monk who lived in Syria after it had been conquered by the Arabs and was the last of the Greek Fathers of the Church to be revered by Roman Catholics. Some people, he said, follow the ancient pagans in asserting that the universe is made of spheres. Others adopt a biblical model of the heavens as a dome over the Earth so that, as the Septuagint says, the Sun travels around the north overnight to get back to where it is supposed to rise. John insisted none of this mattered because, either way, nature obeyed the laws laid down by God. 'Whether it is this way or that,' he wrote, 'all things have been made and established by the divine command, and have the divine will and counsel for a foundation that cannot be moved.'[7] The last thing John wanted was for his more educated readers to think he wasn't up to speed about natural philosophy himself. So, just to make clear he knew exactly what he was talking about, he added a brief but accurate account of what causes an eclipse and the constitution of the heavens according to Aristotle.[8] Later on, he noted the received view of the astronomers that, 'At all events [the Earth] is much smaller than the heaven, and suspended almost like a point in its midst.'[9]

Despite persecution, Christianity became ever more popular in the Roman Empire until Emperor Constantine converted to the new faith himself in 312. As if this wasn't drastic enough, he also moved the imperial capital from Rome to Constantinople (modern Istanbul). It turned out to be an astute move. The western half of the empire fell away in the fifth century after it was occupied by Germanic tribes such as the Goths, Vandals, Franks and Anglo-Saxons. The last emperor in the west was

The Blue Marble, image of Earth taken by the crew of Apollo 17, 1972.

Nicolaus Germanus's 1467 manuscript copy of Ptolemy's world map.

Manichaean *Diagram of the Universe*, Yuan dynasty, 13th–14th century, hanging scroll, paint and gold on silk.

Illuminated page from William of Conches, *Dragmaticon philosophiae*, 14th century.

The first three days of creation, illuminated page in the Sarajevo Haggadah, c. 1350.

Fra Mauro's *Mappa mundi*, 1459–60, hand-drawn parchment mounted on wood.

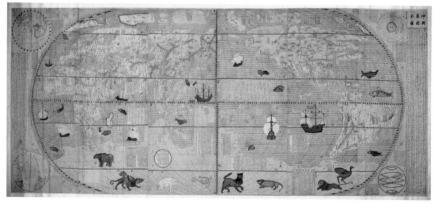

Matteo Ricci's *Map of the Myriad Countries of the World* (*Kunyu Wanguo Quantu*), 1602.

وآول فرشتة يل اوكستة يعني بواده معلق طوررر بقدرة الله تعالي

Islamic world map, illuminated page from Zekeriya Kazvinî, *Acaib-ül Mahlûkat*
(The Wonders of Creation), *c.* 1553.

Cosmic painting of Mount Meru, 19th century, *thangka*,
ink and mineral colours on cotton.

deposed in 476 by the barbarian strongman Odoacer (c. 433–493), who then ruled Italy as king. Modern historians refer to the eastern half of the empire as the Byzantine Empire (after Byzantium, the city that was renamed Constantinople), but it always called itself 'Roman' even after it became culturally and linguistically Greek. Compared to the west, the eastern part of the empire had always been richer and more densely populated. That meant the majority of the world's Christians in late antiquity dwelt in Asia Minor, Syria and Egypt. These communities were so vibrant and numerous that, today, remnants survive – most notably the 5 million or so Copts who live in Egypt.

The Tabernacle

Christians who lived inside the Roman Empire had to obey the strictures of orthodoxy as defined by the emperor. Beyond its borders, other denominations spread across Persia and Central Asia as far as India and China, not to mention Ethiopia. Some of these Christians were religious refugees from Roman tyranny – the losers in the theological controversies that had wracked the early Church. Doctrines on the nature of Jesus Christ, which were denounced as heresy by Constantinople, flourished beyond the reach of the imperial authorities. From the fourth century, the shape of the Earth was one of the issues that divided eastern Christians.

The seed of the dispute grew from the methodologies of two schools of biblical interpretation, usually identified with Alexandria and Antioch, the great city of northern Syria. At a risk of generalizing, the exegetes of Alexandria read their Bibles allegorically, looking for layers of meaning behind the plain sense of the text. Recall how the Jewish Alexandrian Philo pioneered this technique back in the first century AD. In Antioch, they preferred to focus on the stories themselves, which they thought could be taken at face value. The leading light of the School of Antioch in the fourth century was the monk and bishop Diodore of Tarsus (d. *c.* 390).

Little of Diodore's work survives, but we can piece together his doctrines from summaries by later writers. For example, he wrote a trenchant attack on astrology called *Against Fate*, which was the subject of a book review penned in the ninth century by Archbishop Photius of Constantinople (*c.* 815–897). From this, we know Diodore rejected Aristotle's spherical universe saying, according to Photius, 'The sky is not a sphere; it has the shape of a tent or vault. Diodore thinks he can produce scriptural authority for this view, and not only about the shape, but also concerning the rising and setting of the sun.'[10]

We can guess which bits of the Bible Diodore was using as proof texts. When we looked at the Old Testament, we noted that Isaiah 40:22 refers to the heavens as being like a tent. And in the Septuagint, the Book of Ecclesiastes said the Sun travels around the edge of the world at night to return to where it should rise. Greek-speaking Christians considered the Septuagint divinely inspired Holy Writ in its own right, and the Antioch School took it quite literally.

Two of Diodore's disciples were St John Chrysostom (347–407) and Theodore of Mopsuestia (*c.* 350–428). They had been friends since their days together as students at the school of Libanius of Antioch (*c.* 314–*c.* 393), an illustrious pagan teacher of rhetoric. With an education like that, there is no question that John and Theodore were familiar with the evidence for Aristotelian cosmology. Once under Diodore's wing, both men accepted an ascetic lifestyle and devoted themselves to theology. Austerity was a struggle for Theodore in particular, who decided to return to secular life to get married. However, John successfully begged him to come back to the Church, and both men eventually became bishops. John Chrysostom was amazed to find himself elevated to the Archbishopric of Constantinople, making him the most powerful churchman in the Byzantine Empire at a time of great political and religious upheaval.

As members of the Antioch School, John and Theodore expected that everything in the Bible had theological significance and that it meant what it said. They rejected their pagan instruction and embraced Diodore of Tarsus's picture of the world shaped like a tent. And they gleaned reasons for why it had that shape from reading the Bible.[11] The Letter to the Hebrews, an anonymous early Christian text in the New Testament, provided the vital clue.

This epistle compared Jesus to the Jewish High Priest, who ministered in the tabernacle, except that Jesus was the high priest for the whole world, 'who is set on the right hand of the throne of the Majesty in the heavens; A minister of the sanctuary, and of the true tabernacle, which the Lord pitched, and not man' (Hebrews 8:2). The reference to the tabernacle looks back to the Book of Exodus in the Old Testament, which tells of how Moses led the Israelites out of Egypt, across the Red Sea and into the desert on the way to the promised land of Canaan. The tabernacle was a tent that the Israelites carried with them during their wanderings to function as a portable temple. The Book of Exodus set out detailed instructions from God on how it was to be constructed.

According to the Antioch School, the tent in the desert was an avatar for the 'true tabernacle' mentioned in the Letter to the Hebrews, by which it meant the whole world. That is, the universe resembled the tent described in the Book of Exodus. It was shaped like a shipping container: a box with the Earth along the bottom and the sky above. John Chrysostom thought this was a slam-dunk rebuttal of the Aristotelian cosmos. 'Where are they who say that the heaven whirls around?' he demanded. 'Where are they who declare that it is spherical? For both of these notions are overthrown here."[12] Thus, for John, Theodore and Diodore, a flat earth formed part of their biblically consistent world picture. In contrast, Basil of Caesarea and John Damascene admitted that Earth is a sphere, but treated this as an extraneous factoid without theological significance, to be passed over as quickly as possible.

Cosmas on the Cosmos

Early in the fifth century, one of Theodore of Mopsuestia's students, called Nestorius (c. 386–c. 451), became Archbishop of Constantinople. Like some other members of the Antioch School, he was caught up in the schisms that sundered the Christian Church in late antiquity. His mistake was to lambast excessive veneration of the Virgin Mary and suppress her popular title 'Mother of God'. Other clerics objected, and Nestorius was driven into exile before he was formally condemned at the Council of Chalcedon in 451. For good measure, the Church in Constantinople later anathematized Diodore of Tarsus as well. The career of John Chrysostom was just as tumultuous, but he had the good fortune to end up on the side of orthodoxy and is the most respected Father of the Church among Greek Orthodox Christians to this day.

The followers of Nestorius fled eastwards to Persia and the Sassanian Empire. Since their official religion was Zoroastrianism, the Persian shahs spasmodically persecuted Christians as agents of a foreign power but eventually realized many of their Christian subjects loathed the Romans as much as they did. Thus, by the sixth century, much to the disgust of the Zoroastrian priesthood, the shahs tolerated the Nestorian Church of the East and counted its bishops among their confidants.[13] For instance, Mar Aba (c. 490–552), the Patriarch of the Church of the East in the 540s, came from a family of Persian aristocrats.

Born a Zoroastrian, Mar Aba converted to Christianity as an adult. He travelled around the Roman Empire, learning Greek on visits to Constantinople and Alexandria. During Mar Aba's visit to the capital, Emperor Justinian attempted to meet him to discuss the theology of Theodore of Mopsuestia, which Mar Aba was vigorously promoting in the Church of the East.[14] His Zoroastrian upbringing might have made it easier for him to accept Theodore's belief that the tent of the tabernacle was a model of the universe. As we've seen, the world picture taught in the Avesta and other Zoroastrian texts was, like other early models of the universe, based on a flat earth. In contrast, Aristotle's Globe would have seemed like a curious novelty.

Mar Aba failed to catch up with Justinian in Constantinople, but he did make quite a splash in Alexandria. While there, he met a merchant from Antioch called Constantine (*fl. c.* 520– *c.* 550), who had ventured as far as India in search of trading opportunities. Constantine listened to Mar Aba's teaching, which must have included the Antiochene theory about the tabernacle, now most closely associated with Theodore of Mopsuestia.[15] This struck Constantine as a subject he was well equipped to expound, not least because he was widely travelled and knowledgeable about geography. So, when a friend asked him to write an account of the Antiochene world picture, Constantine got straight to work.[16] The resulting book would eventually become history's most notorious defence of the flat earth, known today as the *Christian Topography*. As Constantine didn't sign his work, his name was forgotten, only being rediscovered recently from an Armenian manuscript that quoted him.[17] Instead, the author of the *Christian Topography* acquired the nom de plume Cosmas Indicopleustes, meaning 'the traveller to India'. To avoid confusion, I'll call him by his traditional name of Cosmas, even though he wouldn't have recognized it himself.

The world picture of Cosmas Indicopleustes, illumination from
Topographia christiana, 10th century.

Cosmas had been trading in the Red and Arabian Seas for
25 years before he settled down to work on his book.[18] It's likely
he had amassed a reasonable fortune in that time, giving him
the leisure to write and the wherewithal to commission lavish
illustrations for the *Christian Topography*. Copies of these pictures embellish the three surviving manuscripts of the work,
which date from the ninth to the twelfth century. The book is
an in-depth defence of the world view of the School of Antioch,
explaining that the box-like tabernacle is an accurate model of
the universe.[19] At the same time, Cosmas denied the Aristotelian
theory of the Globe, which, he said, led to such utter absurdities as the existence of the antipodes.[20] He liberally quotes from
members of the Antioch School, such as John Chrysostom, to
support his theories, as well as providing diagrams and pictures.[21] He knew that the Globe was widely accepted among
educated Greeks, for which he blamed the Babylonians for first
suggesting the idea (when, as we've seen, it was really the Greeks
who introduced the Globe to Mesopotamia).[22]

When the *Christian Topography* came out in the 540s, it's fair to say that Cosmas didn't convince the Alexandrian intelligentsia; but at least they didn't ignore him. He even attracted the attention of the greatest Greek philosopher of his day, John Philoponus (*c.* 490–*c.* 570). Although he was a Christian, John had been trained in the Neoplatonic philosophy that dominated the schools of the time. Like Cosmas, he was sceptical of the natural philosophy of Aristotle. But his criticisms of Aristotelianism were subtle, perspicuous and remain valid to this day. For example, he challenged the doctrine that the universe was eternal, showing that this was a logical impossibility, quite apart from conflicting with scripture. He was also the first person to explicitly deny Aristotle's view that heavy objects fall faster than light ones, noting a thousand years before Galileo that a simple experiment disproved this. And he developed a theory of projectiles to explain why, contrary to Aristotle, objects continue to move after the force moving them has been taken away.[23] However, on one point, John Philoponus and Aristotle were in complete agreement: the Earth is a sphere.

In his massive commentary on the creation story in Genesis, probably written in the 550s, John had Cosmas in his sights. While he didn't deign to mention Cosmas directly (instead naming Theodore of Mopsuestia as his archetypical flat-earther) he did include several direct allusions to the *Christian Topography*.[24] The challenge for John was the lack of clear evidence for the Globe in scripture. Although he revered the Bible as the word of God, he couldn't use it to argue that the Earth is a sphere and was reduced to insisting that it doesn't teach astronomy at all. Instead, he pointed to empirical observations to show the universe is spherical.[25] He noted the evidence from eclipses, the movements of the stars and the rising and setting Sun.

As for Cosmas, his reading of scripture was naive. But, because the biblical authors did assume the Earth was flat, he had an easier time arguing they supported his view. Even if the theory that the universe resembled the tabernacle was rather contrived, he could at least find evidence for it in the Bible.

Perhaps John's diatribe had the effect of suppressing the spread of the Antiochene world picture. There is plenty of evidence that variants of it were prevalent in eastern churches during the sixth century.[26] But we hear much less about it after this period, among both Syrian and orthodox Christians. It seems likely that, among the educated at least, the Globe gradually became more widely accepted. For example, Ananias of Shirak (c. 600–670), an Armenian Christian scholar of the seventh century, was fully convinced by Aristotelian cosmology and could accurately explain the cause of eclipses.[27] At the same time, Christians in Syria were teaching their students about astronomical instruments in a way that took the Globe for granted.[28] In the ninth century, Photius of Constantinople was even less impressed by the *Christian Topography* than he had been by Diodore's *On Fate*. Its style, he said, was 'low and does not even follow ordinary syntax'; there was much that was 'historically implausible' and the author could fairly be regarded as 'more devoted to myth than truth'.[29]

Nonetheless, it is an exaggeration to say that 'the influence of Cosmas's blundering efforts on the Middle Ages was virtually nil,' as one twentieth-century historian has put it.[30] Admittedly, no one translated the *Christian Topography* into Latin until the eighteenth century, so it was unknown in the West until modern times. But it was far from forgotten in the Byzantine Empire. The three complete surviving manuscripts are sumptuously illustrated with reproductions of the pictures that Cosmas himself had originally commissioned. Whoever ordered these manuscripts felt that the *Christian Topography* was worth lavishing a serious amount of money on. Around the same time, Byzantine collections of quotations from the Church

Fathers often featured comments on Genesis by members of the School of Antioch.[31] One modern scholar went so far as to say, 'There can be little doubt that the Antiochene conception of the universe, as exemplified by Cosmas, reflected the views of the Byzantines on this subject.'[32] That takes things too far, but there were certainly some strange cosmological ideas circulating among medieval Greeks. For example, an anonymous Byzantine writer thought that eclipses were caused by a dark star, not unlike the planet Rahu in traditional Hindu astrology.[33] Even though the Aristotelian world picture became steadily more prevalent, the ideas of Cosmas and the Antioch School probably hung on in monasteries for as long as the Byzantine Empire existed.

Throughout their history, the Byzantines were under attack. They had to defend themselves from tribes to their north and Catholic crusaders to their west. But their most trenchant foes came from the east. As we've seen, they fought Sassanian Persia to a standstill in the early seventh century, leaving both empires exhausted. This meant they could not defend themselves from the Muslim Arabs who overthrew the Sassanid dynasty. After defeating the Byzantines at the Battle of Yarmouk in 636, the Arabs seized the rich provinces of Syria and Egypt. Seven hundred years later, in 1453, Ottoman Turks took Constantinople, finally snuffing out the last remnant of the Roman Empire.

15

Islam: 'The Earth laid out like a carpet'

Shortly after lunch, on Friday, 30 July 762, the Caliph al-Mansur (714–775) founded his new capital on the banks of the Tigris. In a show of optimism, he named it the 'City of Peace', but posterity calls it Baghdad. We know the precise time the ground was broken for the building of its circular walls because al-Mansur commissioned a team of astrologers to calculate the most propitious moment. The team included a Persian, al-Fazari (*fl. c.* 760–*c.* 780), and Mashallah (*fl. c.* 760–*c.* 815), a Jew whose works on astrology would become famous in medieval Europe.[1]

Al-Mansur was the second of the Abbasid caliphs. His father had overthrown the Umayyads, the first Muslim dynasty, which had presided over the rapid initial expansion of Islam. Although the Abbasids were Arabs and distant relatives of the Prophet Muhammad (*c.* 570–632), they had supplanted the Umayyads thanks to an army from Persia. The fall of the Sassanian Empire had left the Persian elite dispossessed and resentful, so when the Abbasids rebelled, they were more than happy to help overthrow the oppressive Umayyads.

While the Abbasid revolution superficially looks like the replacement of one Arab dynasty with the other, in reality it was more like a reverse takeover by the Persians of their Arab conquerors. While al-Mansur had prudently arranged for the murder of his most successful Persian general, he still needed to propitiate the rest of his Iranian allies. Siting his new capital just north of the old Sassanian city of Ctesiphon, not to mention recruiting Persian bureaucrats to run the caliphate for him, helped build legitimacy in their eyes. Eventually, the Abbasid caliphs became figureheads for the Persian viziers who ran the government.[2] Over the long term, the replacement of the Umayyads, whose capital was Damascus, with the Abbasids of Baghdad led to the loss of Arab dominance east of the River Tigris. Although the Persians converted to Islam, and Zoroastrianism went into decline, they refused to accept Arab hegemony and preserved their language. Today, Arabic is the lingua franca from Morocco to Iraq, but Iranians speak Persian and boast a stupendous poetic tradition in that language.

The Translation Movement

The Umayyads had kept their Arab soldiers separate from the people they were conquering, building garrison cities outside the main population centres. The Abbasids recognized that they needed to be more cosmopolitan if they were going to hold the caliphate together. Unlike the Umayyads, they encouraged non-Arabs to convert to Islam while also seeking to draw the leaders of other faiths, such as Jews and Christians, into their orbit. For example, Timothy I (727–823), who occupied the office of patriarch of the Church of the East once held by Mar Aba, was required to base himself in Baghdad, where he engaged the caliph in polite debate about the merits of their respective creeds. And when the caliph wanted an Arabic translation of one of Aristotle's works on logic, he expected Timothy to provide it.[3] While dialectic was rendered directly into Arabic from the Greek, Aristotle's cosmology travelled in a more roundabout fashion. In fact, it's likely that al-Mansur first learnt about the theory of the Globe a decade after the foundation of Baghdad, not from Greek or Persian sources but from India.

Once the power of the Abbasids was secure, neighbouring countries sent envoys to the capital to check them out. So it was that, around 773, an embassy arrived from India at the newly completed city of Baghdad, where the Caliph al-Mansur was holding court. We don't know from which of the various Indian kingdoms the embassy came, but it is likely to have been among the successors to the Gupta Empire that had collapsed in the sixth century. One of the ambassadors was carrying a book on mathematical astronomy, written in Sanskrit, which drew the attention of the caliph. At the time, the reputation of Indian astronomy was at an all-time high. As we'll see, there were Buddhist astronomers working for the Chinese emperor, something that al-Mansur may well have been aware of since he'd been sending embassies to China.[4] There was already a border between the caliphate and China in Central Asia, where the two powers had fought the Battle of Talas back in 751. Prisoners captured by the Muslim side had introduced papermaking to the Middle East, so the production of books was a good deal cheaper than before.[5]

On the orders of the caliph, al-Fazari, one of his court astrologers, arranged for the Sanskrit book to be translated into Arabic. It was called the *Zij al-Sindhind*: the earliest book of mathematical astronomy in Arabic.[6] The core of the work consisted of astronomical tables of the sort that astrologers needed to determine the positions of the planets and produce horoscopes. The word *zij* comes from the Persian for 'thread', since the criss-crossed lines in the tables resembled the warp and weft of a tapestry.[7]

Neither the Sanskrit original nor the Arabic translation of the *Zij al-Sindhind* survive, but we can get an idea of the contents from later quotations. It appears to be based on an astronomy textbook written in India about a century after the death of Aryabhata. According to the Muslim natural philosopher

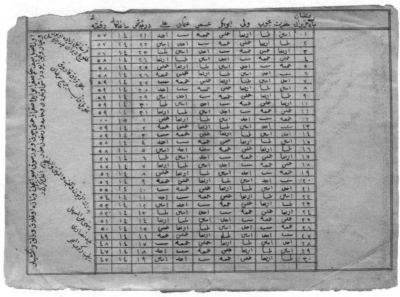

Islamic astronomical book (*zij*).

al-Biruni (973–*c.* 1050), al-Fazari found a figure for the circumference of the Earth of about 40,000 kilometres (25,000 mi.) from his Indian source.[8]

There were already books on astronomy in Greek and Syriac available in Baghdad at the time of the Indian embassy, and some of these would have alluded to the Globe. After all, Mesopotamia hosted a substantial Greek population and their schools taught Aristotelian philosophy. However, these Greek works only started to be translated into Arabic a few years later. In any case, the importance that Arab historians attach to the *Zij al-Sindhind* of al-Fazari shows they considered it the beginning of the Muslim achievement in astronomy.[9]

The translation of Indian, Persian and Greek works into Arabic in Abbasid Baghdad was one of the great events of intellectual history. Caliph al-Mansur began the programme after the foundation of the city in the 760s. He was particularly interested in Persian guides on how to rule an empire, since he had effectively become the heir to the Sassanids, for whom these books were originally composed.[10] The resulting collection was kept in a library called the House of Wisdom, which may have been little more than a room in the palace. However, its evocative name has led to a myth that the House of Wisdom was akin to a university, or at least the centre of the Abbasid translation movement. Neither of these things are true.[11] Some of the caliphs did grant audiences to their menagerie of scholars and even questioned them on their work. But the idea that Baghdad harboured an institution similar to a modern university or the (equally mythical) Great Library of Alexandria is not supported by the evidence.

The caliph most associated with the translation movement is al-Mansur's grandson al-Mamun (786–833). Al-Mamun's father had decreed that his younger son, al-Amin (787–813), would succeed him as caliph. He appointed al-Mamun governor of the key province of Khorosan, roughly coterminous with northern Iran and Afghanistan. However, the relationship between the brothers rapidly broke down following their father's death and, after a fratricidal civil war, al-Mamun became caliph in 813. The translation movement was part of his effort to centralize and unite the disparate parts of the caliphate. By rendering Persian and Greek scholarship into Arabic, he was asserting that his realm was the rightful successor to the Sassanian and Roman empires.[12]

The translation movement was utilitarian, biased towards practical material like medical textbooks. Astronomy was useful primarily because astrologers needed it to carry out their work. Islamic scholarship fully assimilated the spherical astronomy of Ptolemy in 805 when his *Mathematical Synthesis* was rendered into Arabic.[13] It became known as *Almagest*, meaning 'the Great', a title it retained when it was transmitted to Europe during the twelfth century and continues to enjoy today. Aristotle's *On the Heavens* wasn't translated until around 850, after Caliph al-Mamun had died, but contained little about the Globe not already known from Ptolemy. Ninth-century Muslim philosophers working in Baghdad, such as the Arab al-Kindi (801–873) and the Persian al-Khwarizmi (*c.* 780–*c.* 850), incorporated spherical astronomy into their research. In a short letter on the subject, al-Kindi used geometry to show why, if Aristotle was correct to say that all heavy objects fell to the centre of the universe, the Earth would have to be a sphere located there.[14] Al-Khwarizmi is one of the best-known Muslim mathematicians in the West because his book on algebra was translated into Latin and formed the basis of arithmetical education in medieval Europe. Over the next few centuries, astronomers writing in Arabic continued to refine Ptolemy's work, attempting to find more elegant solutions than his to explain the motions of the planets.[15]

During the Abbasid era, Muslims called the learning from Indian, Persian and Greek sources 'foreign sciences'. This wasn't derogatory. It was just a statement of fact that distinguished them from subjects like law and the interpretation of scripture.[16] When the translation movement began in the late eighth century, the study of the Koran was already a mature discipline. And since the Koran was the product of a very different environment from multicultural Baghdad, its world picture didn't cohere with the cosmology transmitted by the foreign sciences of Indian and Greek astronomy.

The Koran

The Koran, the record of the teaching recited by Muhammad, which Muslims believe to be the word of God, initially circulated as separate chapters, or *surahs*, before being collated under the Umayyad Caliph Uthman (*c.* 576–656) around AD 650. Since then, the Koran's text and organization have remained almost immutable. To be a Muslim obviously means to accept the Koran was revealed to Muhammad by God. Most secular historians have concluded that the text we have today is a reliable record of his preaching, not least because its content so closely reflects the milieu of early seventh-century Arabia.[17]

According to the Koran, the universe is the creation of God, encompassed by 'the heavens and the Earth', a phrase it uses over two hundred times. The heavens are plural because there are seven of them, arranged in a stack: 'He created seven heavens, one above the other.'[18] Above, there are waters, echoing the waters above the firmament in the Book of Genesis, surmounted by the throne of God. The lowest of the seven celestial tiers is the sky that we can see above us. It is a solid protective roof maintained in existence by the continuous action of God.[19] The Earth is 'laid out flat', spread like a carpet or a bed, studded with mountains to keep it stable: 'We spread out the Earth and set upon it immovable mountains.'[20] There's a suggestion that there are seven earths to match the seven heavens, although this idea is not fully developed in the Koran itself.[21]

As we'd expect, the world picture in the Koran reflects traditional beliefs of the Arab tribes. Like the Bible, it refers to heaven and earth in the context of God's power and might, praising his creation and sustenance of the universe. Any details it provides about the structure of the world are incidental and irrelevant to its core themes. Introducing novel ideas about the shape of the world would have risked muddying the religious message Muhammad sought to impart.

While its basic world picture is reasonably clear, there is much that the Koran is not explicit about. For example, it doesn't tell us whether the Earth is a disc or a square.[22] Nor do we learn much about the edges of the world. We've seen how seafaring civilizations, like the Greeks, surrounded the Earth with a great ocean, while landlocked Persia imagined that the borders of the Earth were mountainous. The Koran rejects both of these. The Arabs traded across the desert, while both Mecca and Medina, the cities associated with Muhammad, are inland. So, we would not expect the edge of their world to be an ocean. Indeed, in a story about Alexander the Great, recounted by the Koran, we follow the conqueror to the Far East and Far West where he witnesses the rising and the setting of the Sun. Each evening, the Sun falls into a muddy pool rather than the sea.[23] The Koran states both the Moon and Sun travel in circular orbits through the atmosphere under the lowest heaven and then, presumably, under the Earth at night.[24] It doesn't explicitly describe the vault of the sky.

Muhammad was a merchant and if his trading missions had taken him to Syria, as seems very likely, he would have encountered Christians from the Church of the East who thought the universe was like the biblical tabernacle. This might have caused Muhammad to contemplate the form of the Kaaba, the cuboid building that was the central shrine of Mecca and is now surrounded by the Great Mosque.[25] Perhaps it was the Kaaba rather than the tabernacle that reflected the shape of the world.

The Sayings of the Prophet and His Companions

The Koran is the ultimate authority in Islam, but there are other texts said to derive from the words of Muhammad and his close companions. These sayings were remembered by his followers and transmitted orally until they were written down after a couple of centuries. Each saying, or *hadith* in Arabic, is a pithy fragment of a conversation in which the Prophet usually answers a question from one of his followers. Unfortunately, it became evident that a great many of Muhammad's alleged utterances had been invented later on, to win an argument or prove a point.

Their numbers proliferated, and blatant inconsistencies crept into the canon. Muslim scholars were alive to this issue and went to great efforts to authenticate the sayings by verifying the chain of oral transmission.[26] Many were declared 'weak' and so treated with scepticism. By AD 900, specialist researchers had winnowed down the thousands of sayings into six overlapping canonical compilations that, according to Muslim consensus, enjoy a high level of reliability.[27]

The sayings are an invaluable record of the debates and thinking of Muslim scholars in the earliest years of Islam. In particular, they reflect the intellectual environment before the translations of Greek astronomy and philosophy had taken hold. In this respect, a weak *hadith*, while not reflecting the words of Muhammad himself, is still evidence of Muslim opinion in the years before 900.

A *hadith* elucidates the meaning of the Koran or addresses an issue that isn't explicitly covered in scripture. While the majority of the sayings deal with legal issues, there are some that mention the constitution of the universe. Building on the words of the Koran, many sayings state there are seven earths arranged in layers, mirroring the seven heavens. For example, a canonical *hadith* states, 'Whoever takes a piece of the land of others unjustly, he will sink down through the seven earths on the Day of Resurrection.'[28] In contrast, a Persian scholar called al-Tabari (829–923) cited a *hadith* that denied the earths are on top of each other. 'There are seven earths that are flat and islands,' he quotes. 'Between each two earths, there is an ocean.'[29] This might be influenced by the world pictures of the Indian Puranas or Zoroastrian Avesta, with their continents separated by seas.

In the fifteenth century, al-Suyuti (c. 1445–1505), a polymath from Cairo, compiled many sayings on traditional astronomy from the *hadith* collections and religious experts.[30] Al-Suyuti's collection contains a bewildering variety of ideas about what the world looked like, as well as evidence of cross-pollination between different traditions. Some of the sayings he recorded affirm that there were seven earths, one on top of the other, each separated by a distance that would take five hundred years to traverse. The seven heavens over the earths were separated by a similar interval. Above the heavens was a great ocean, over which eight enormous mountain goats supported the throne of God. Another saying omitted the goats but stated that the Earth rested atop a fish, which caused tremors when it stirred. A more traditional idea was of a mountain that surrounded the Earth and rooted it to the rock on which it rests.[31] The diversity of these traditions shows that, while there was a consensus on the basic schema of seven heavens and seven earths, Muslims had latitude to develop and embellish this model as they saw fit.

The Koranic world picture was paradigmatic among a large segment of the Islamic population for many centuries, even after knowledge of the Globe had become widespread.[32] However, it was not a point of doctrine that faithful Muslims were expected to adhere to. As a very popular commentary on the Koran from the fifteenth century, called the *Tafsir al-Jalalayn*, noted,

As for His words [in scripture], 'laid out flat',[33] this on a literal reading suggests that the Earth is flat, which is the opinion of most of the scholars of the revealed Law, and not a sphere as astronomers have it, even if this latter view does not contradict any of the pillars of the Law.[34]

Even Muslims who opposed aspects of Greek philosophy gradually became habituated to Aristotelian cosmology. One example was Abu Hamid al-Ghazali (1058–1111), perhaps the most famous of all Islamic theologians, whose genius made him a powerful foe of thinkers who were enamoured by the foreign sciences translated into Arabic. He started his career teaching in the leading theological college in Baghdad, where he wrote *The Incoherence of the Philosophers*, a trenchant attack on Aristotelians who sought to meddle in religious affairs. His main target was Avicenna (980–1037), the brilliant thinker whose work forms the foundation of so much Arabic philosophy. Al-Ghazali himself was steeped in the Greek philosophical tradition, which is precisely why he was such a dangerous opponent for Avicenna's followers. He successfully fought them on their own ground.

In *The Incoherence of the Philosophers*, al-Ghazali analysed twenty theses, largely derived from the thought of Avicenna, such as the eternity of the world and whether the heavens are alive, and showed they are either false or not proven. Al-Ghazali didn't set out doctrines of his own, at least not in this book, but rather punctured the pretensions of those who taught that philosophy had all the answers. His most famous refutation is the seventeenth, in which he challenged the axiom of cause and effect, showing that it isn't possible to conclusively demonstrate that any physical law will always hold true.

Al-Ghazali was careful in his choice of the theses he attacked, so it is revealing that the theory of the Globe was not among them. Despite the traditional world picture in the Koran, he took the basics of the Aristotelian universe for granted. He assumed the universe is a sphere and even gave Aristotle's *On the Heavens* a name check, saying there was no need to argue about it.[35] Theoretical and practical science was not necessarily contrary to divine law, he said, even if philosophers did tend to lead themselves astray. Muslim commentators should accept that God could have made the world in any way he wanted to. Aristotle might be right to say the Earth is a sphere in the middle of a spherical universe, but he was wrong to say it had to be that way. God, in his wisdom, could have built the world in any shape he pleased.[36] Likewise, the faithful didn't have to worry about physical causes, such as how things stayed attached to a round earth, since God had arranged that they should do so.

In his autobiographical work *The Deliverance from Error*, al-Ghazali goes further in his admonitions of those who would reject science. He warned that a pious Muslim, who ignored the naturalistic explanations for spectacles such as eclipses, might encourage people to disdain Islam. Anyone who rejected astronomy outright, thinking that they are defending the faith, risked doing great harm.[37]

The view of al-Ghazali and others that the shape of the Earth was not part of religious doctrine didn't overthrow the Koranic world picture overnight, but it did take the heat out of the argument. Muslim philosophers and mathematicians enjoyed a consensus that the Earth was spherical from the ninth century. Religious elders slowly came around to this view. They were not naive and were comfortable reading the Koran less literally, if that was necessary to glean what they took to be its true meaning.[38] Without much drama, and over many centuries, they reinterpreted passages from scripture in a way that was consistent with the Globe.

Today, there are still occasional reports of Muslims who subscribe to a traditional world picture. On 12 February 1995, the *New York Times* reported that the Grand Mufti of Saudi Arabia, Abd al-Aziz ibn Baz (1910–1999), had issued a fatwa a couple of years previously ruling that anyone who said the Earth is round is an unbeliever who should be killed.[39] The report was untrue. However, back in 1966, Sheikh Bin Baz had asserted that it was heretical to state the Sun is stationary. Exactly what he thought about the shape of the Earth is unclear. The journalist Robert Lacey reported that the sheikh personally supposed it was flat. Nonetheless, he didn't think this was a matter on which he would find fault with his fellow Muslims.[40] Just like the author of the *Tafsir al-Jalalayn* in the fifteenth century, Sheikh Bin Baz acknowledged that men of science said the Earth was a sphere, but he didn't hold to that himself. People were free to decide either way. However, some commentators have denied he thought the Earth is flat, pointing to his writings where he explained that the language of the Koran was based on everyday appearances. Flatness is an impression caused by the sheer size of the Earth.

Sheikh Bin Baz was born into a family of cloth merchants in Riyadh. His father died when he was young and by the age of twenty he had lost his sight due to an infection. Overcoming this disadvantage, he excelled in his studies of Islamic law and became an exceptional scholar. He would have known the Koran by heart (not an unusual accomplishment among devout Muslims) and been intimately familiar with the *hadith* collections and other classical sources of Islamic law. Since he was blind by the time Riyadh was exposed to Westernization during the oil boom, he would never have seen the ubiquity of globes and photographs of Earth from space so familiar today. Intellectually, he grasped the theory of the Globe and could explain why it was compatible with a plain reading of scripture. But perhaps, inside, he never quite believed it.

Which Way to Mecca?

The shape of the Earth may not have been a matter of doctrine, but it did have an important religious implication. Wherever they are, Muslims need to know the direction of Mecca. The Koran stipulates that the faithful should pray facing the Kaaba, the ancient shrine at the centre of the holy city. When constructing a mosque, the architects would orientate it accordingly and include a niche in the relevant wall to mark which way to face. Unfortunately, given that the Earth is a ball, it is a ticklish

The Kaaba, Mecca.

problem to determine exactly which bearing to use. You need to know your latitude and longitude as well as that of Mecca itself. Then you have to apply some spherical trigonometry.[41] Nowadays, you can download an app on your phone to deal with all of this, but early Muslims had to make do without accurate maps or advanced mathematics. As a result, some ancient mosques aren't aligned quite as might be expected.[42]

Travellers consulted charts to tell them the direction of Mecca in various cities. They assumed that the *Kaaba* was at the centre of the Earth, with other cities dotted around it. Many manuscripts survive showing diagrams of this arrangement. Because these charts assumed the Earth is a plane, it didn't matter how far away each city was, only the direction between it and Mecca. To figure out where to face when praying, you would examine one of these guides, which showed the astronomical sightings that would allow you to locate the Kaaba.[43] These aids, which illustrated where a particular constellation rose above the horizon, had been essential to navigation in the desert long before

Muhammad's ministry and would have been much more familiar to ordinary people than the mathematical methods favoured by astronomers schooled in the Aristotelian tradition.

Because the traditional methods took no account of the Globe, they were not accurate at long distances. By the ninth century, Muslim mathematicians had mastered the techniques required to compensate for the curvature of the Earth, but they required the co-ordinates of the Islamic world's cities to carry out the calculations.[44] As we saw, Ptolemy of Alexandria had produced a list of latitudes and longitudes for a litany of places in his book *Geography*. The astronomers of Baghdad had access to this material, but it didn't solve the problem. Ptolemy's gazetteer omitted numerous locations they were interested in – not least the city of Mecca itself.

The sequence of events that led to Muslim scholars compiling their own list of co-ordinates isn't clear. Caliph al-Mamun commissioned a map of his empire that several writers later mentioned, although no copy survives. He also dispatched two expeditions to determine the size of the Earth. They both paced out the distance of one degree of latitude, which they determined from the position of the stars, giving figures for the Earth's circumference of between 38,500 and 40,000 kilometres (24,000 and 25,000 mi.), consistent with the modern figure. Finally, the caliph ordered the mathematician al-Khwarizmi to collate the latitudes and longitudes of some five hundred locations. Unlike al-Mamun's map, this table of co-ordinates survives.[45] The combination of al-Khwarizmi's table and the estimates for the size of the Earth would have allowed the caliph's cartographers to create a map of his domains that showed both locations and distances, at least as long as the data they incorporated was accurate. It's easy to see why such a chart would have had important political and military applications, although it is not clear what it was used for.

Islam's sacred geography, manuscript, Yemen, 1356–7.

There is no reason to think that al-Mamun sponsored his cartographical project with an eye to properly orientating mosques, although it would have been a handy by-product. However, while the ingredients necessary to precisely calculate the direction of the Kaaba were all in place by the ninth century, the method doesn't seem to have been widely used. Guides to finding Mecca that utilize latitudes and longitudes do exist from the thirteenth century, but they are rare.[46] As with cosmology, traditional and spherical direction-finding methods coexisted for centuries before the Globe finally became the dominant theory in the eighteenth and nineteenth centuries.[47]

16

Later Judaism: 'The wise men of the nations have defeated the wise men of Israel'

For Jewish communities in the Middle East and Mediterranean basin, the substitution of the Byzantine or Persian empires for the Islamic Caliphate was of little moment. In Spain, where the Visigoth kings had celebrated their union with the Catholic Church by enacting antisemitic legislation, Muslim rule was a distinct improvement.

By this time, most Jews accepted the authority of the rabbis, for whom the language of worship and law was Hebrew. Greek-speaking Judaism, which had produced the Septuagint version of the Bible and Philo of Alexandria, had largely died out. The Talmud and the Tanakh were the foundation of religious culture. So, when the Abbasid caliphs first sponsored the translation of Greek philosophy and science in the eighth and ninth centuries, it had little impact on Jewish scholarship. For example, the *Chapters of Rabbi Eliezer the Great*, a biblical commentary composed around AD 800, describes the sky as like an upended tub or tent over the Earth, which 'is spread upon the waters like a ship which floats in the midst of the sea'.[1] Later, the *Chapters* echo the Book of Enoch's description of the Sun rising and setting through gates at the edge of the world. Even the growing appeal of astrology among Jewish people doesn't seem to have affected their world picture.

The Jewish Rediscovery of Greek Thought

During the ninth century, as Greek thought became better established among Muslim intellectuals, it was natural that Jewish scholars would start to take notice. The man who did most to show how it was possible to exploit philosophy to serve the Jewish faith was Saadia Gaon (882–942). The word 'gaon' in Saadia's moniker signalled that he was the head of a religious school, a post that Saadia held at the city of Sura in Mesopotamia.[2] He had been born in Egypt but moved to Sura at the request of the leader of the local Jewish community. Saadia had drawn upon his knowledge of astronomy to heal a schism over the date of the new moon. This was vitally important, since the lunar calendar determined when major religious festivals were celebrated.

Saadia had seemed to be the ideal man for the post at Sura, but he was not as tractable as his patron had hoped. Within a few years, his incorruptibility triggered a row after he refused to endorse a questionable legal decision. Expelled from Sura, he fled to Baghdad, where he was able to devote himself to study. Drawing on the work of Muslim philosophers, he argued that a proper understanding of religion required the application of reason as well as faith. His most influential work, written while he was living in Baghdad, was the *Book of Beliefs and Opinions* – the first example of rabbinic philosophical theology.[3]

Saadia's authority as a rabbi and a scholar meant he could effect a major realignment of Jewish thought. As Muslim theologians had been doing for some time, he used philosophy to validate religious doctrine and to challenge heresy, setting out the criteria for when reason could legitimately help Jews to live a good and dutiful life. Ceremonial rules were instructions from God, he explained, so following them was a matter of faith. In contrast, the ethical precepts in the Bible and the Talmud could be derived from rational analysis. Furthermore, reason was a powerful weapon against religious error.[4] Like John Philoponus, Saadia used it to rebut theories that the universe was eternal rather than being created from nothing by God.

He didn't directly address the Globe in the *Book of Beliefs and Opinions* and only mentioned it in passing in his commentary on Genesis, where he noted that the Earth is spherical with only part of the Northern Hemisphere habitable.[5] His most substantive remarks on the issue occur in notes he wrote in Egypt as a young man, commenting on a mysterious work called the *Book of Formation*. Today, the *Book of Formation* is considered as an early source for the Kabbalah, the distinctive form of numerological mysticism that would become prevalent among Jews in the late Middle Ages.

Saadia translated the *Book of Formation* from Hebrew into Arabic and added his own, much longer, commentary. Since the original text was concerned with the mystical significance of numbers, it might seem odd that Saadia drew implications from it about the shape of the world. Be that as it may, he interpreted the *Book of Formation* as supporting the traditional world picture whereby the sky was a vault over a flat earth – 'as a roof covers a house'.[6] He took this to be the view of some earlier rabbis, including Joshua ben Hananiah (who had met the emperor Hadrian and flummoxed the Athenian philosophers) and the author of the *Chapters of Eliezer* quoted above. All these authorities followed the Book of Enoch in saying the Sun never sank beneath the Earth. Rather, at night, it hid behind the opaque firmament as it travelled back to the place it should rise.

Rejecting the received view, Saadia espoused Aristotelian cosmology: heaven and Earth were spherical, with the former a point at the centre of the latter. He explained that the Sun sank under the Earth at night, which blocks its light until it rises the next day. To support his contentions, he took advantage of the multiplicity of previous opinions among the rabbis to make out there was a precedent for his novel views. Suspicions that people might have about innovative ideas could be overcome by showing they weren't new at all. For example, he quoted the passage in the Talmud where Judah ha-Nasi accepted that non-Jewish scholars were probably right about the Sun passing beneath the Earth.[7] Admittedly, Judah ha-Nasi had never said the Earth was a sphere, but his concession that pagan cosmology was occasionally correct made it much easier for Saadia to argue in favour of the Globe.

Other Jewish writers started to adopt Aristotelian cosmology shortly thereafter. One of the earliest was a certain Dunash ibn Tamim (*c.* 890–960). He composed his own commentary on the *Book of Formation* and a handbook on astronomy called *The Configuration of the Orb*, which he dedicated to Egypt's Muslim ruler.[8] As for Saadia Gaon, with his reputation for sagacity and holiness now unimpeachable, he returned to the religious school in Sura for the last five years of his life. By demonstrating the utility of philosophy as a tool to strengthen the faith, he inaugurated a new direction in Jewish thought. Aristotle became a potential ally in the battle for religious truth, who didn't need to be disparaged just because he was a pagan. Unfortunately, there was no denying that a lot of what he had said was difficult to reconcile with the Bible or the Talmud. Two centuries after Saadia's death, medieval Judaism's most famous thinker, Moses Maimonides (1138–1204), met this challenge head on.

Moses Maimonides

Maimonides was born in Cordoba, then a great intellectual centre of Islamic Spain. Unfortunately, while he was still a child, the city came under the rule of an intolerant regime that threatened Jews and Christians with death if they failed to convert.

Maimonides's family fled, eventually settling in Cairo. By then, Maimonides himself was already a noted teacher and the family enjoyed some prestige, while his younger brother became a merchant, sailing to India in search of spices. Sadly, the voyage was hazardous and he drowned at sea. Maimonides was devastated but rebuilt the family's fortunes. He rose to become a leader of Cairo's Jewish community and physician to the sultan. Legend has it he died of overwork as an old man.

Whereas Saadia had largely relied on second-hand knowledge of Greek philosophy from Muslim sources, Maimonides was deeply immersed in the original texts, albeit in Arabic translation. His cosmology followed the spherical model of Aristotle very closely. As he put it in his *Foundation of the Torah:* 'All of these spheres which encircle the universe, are round as a globe, and the Earth is suspended in the centre.' These spheres were like the skins of an onion, with no space between them. Maimonides was only interested in summarizing astronomy but mentioned there was plenty more to read on the 'the science of cycles and constellations, concerning which the Greek scholars wrote many books'.[9]

Although he accepted much Aristotelian doctrine, as an observant Jew, Maimonides was uncomfortable concluding the Hebrew prophets could be wrong. Instead, he insisted that contradictions between philosophy and the Bible were illusory. The trouble was, he said, that 'wicked barbarians have deprived us of our possessions, put an end to our science and literature, and killed our wise men; we have become ignorant.' In other words, the disasters that had befallen the Jewish people had robbed them of the ability to genuinely understand their heritage, a problem only worsened by them living amid a boorish population. 'We are inclined to consider these philosophical opinions as foreign to our own religion, just as uneducated persons find them foreign to their own notions,' he lamented. 'But, in fact, it is not so.'[10]

To help students harmonize Aristotelian philosophy with Judaism, Maimonides wrote his most famous work, *The Guide for the Perplexed*. He used several techniques to reconcile scripture and reason, including careful analysis of the text of the Bible to show that it can be interpreted in a way that is consistent with philosophy. He found the authority of the Talmud harder to uphold but was helped by its inclusion of disagreements among the rabbis it quoted. For example, to justify accepting aspects of Aristotelian cosmology, Maimonides noted the same passage from the Talmud that Saadia had done, where Judah ha-Nasi allowed that pagan philosophers may sometimes have been right on astronomical questions. Maimonides paraphrased Judah ha-Nasi slightly as saying that 'the wise men of the other nations have defeated the wise men of Israel' on this question.[11]

When it came to any remaining inconsistencies between Aristotle's cosmology and the traditional world picture, Maimonides was more relaxed. On matters of speculation, he said that everyone was entitled to accept whatever conclusion they thought had been demonstrated most convincingly.[12] With other issues, like the distance between the Earth and heaven, the Talmud might only reflect the best knowledge available at the time.[13] And some things were simply unknowable, such as the precise order of the planets' orbits outwards from the Earth.[14]

These opinions were controversial enough among his fellow Jews while Maimonides was alive. After his death, the dispute over his legacy raged as his writings spread into Europe. The rabbis of northern France, renowned for their knowledge of the Talmud, were displeased by his promotion of reason and philosophy as weapons in the service of religion. Their leading light had been Jacob ben Meir (d. 1171), who had no time for rationalism and subscribed to the traditional world picture of the Bible. For example, he said the Sun could not pass beneath the Earth and he probably also held that the world was flat with a heavenly dome over the top of it.[15]

Rabbinic suspicion of Maimonides could only have been exacerbated by the way that Aristotelian philosophy was sweeping through Christian intellectual circles at the same time. French Jews knew how precarious their existence was. The memories of mass murders by wannabe crusaders on the Rhine in 1096 were a warning that their Christian hosts could turn on them at any moment. Indeed, it wasn't long before the Catholic authorities got involved in the squabble over Maimonides. They examined his books, declared them heretical and had them burned in 1232. This hostile intervention shocked the Jews who, recognizing who their common enemy was, buried the hatchet and acknowledged Maimonides's devotion to God and the Bible, even if they continued to disagree about his conclusions.[16]

Nonetheless, in spite of what the more conservative rabbis thought about it, Aristotelianism slowly infiltrated European Judaism, starting from the southwest. Spanish Jews had long been familiar with philosophical thought because of the strength of Aristotelianism in the Iberian Peninsula, personified by the Muslim philosopher Averroes (1126–1198). For example, the Zohar, a primary text of the Kabbalah composed in thirteenth-century Spain, referred to the Earth as an orb. It further explained that the seven heavens were arranged spherically and, as Maimonides had done, compared this to the skins of an onion.[17]

Similarly, the Sarajevo Haggadah, an illuminated manuscript of readings for Passover, created in Spain in around 1350, depicts an unambiguously spherical earth in its illuminations of the Creation. This unique book combines Greek and Jewish imagery, taking both as sources of authority, and incorporates a round earth into a universe with a vaulted roof that represents the tabernacle.[18]

While they were slower to welcome the entry of philosophy into religious law, French and German Jews gradually realized its value. For example, Gersonides (1288–1344), a provocative rabbi from the south of France, was well aware of the Globe and attempted to improve Ptolemy's model of planetary motion. Facing the inevitable, many rabbis concluded that since the shape of the Earth didn't touch upon property, ritual purity or anything that might be considered a capital offence, they didn't need to provide a definitive ruling on it.[19] Still, even as late as the sixteenth century, there were a few who rejected any findings of science – ancient or modern – that conflicted with the picture of the world assumed by the scriptures and Talmud.[20]

The Order of Preachers, better known as the Dominicans after their founder, St Dominic, had been at the vanguard of the Christian attack on Moses Maimonides. Yet, the order was also among the most enthusiastic proponents of the reconciliation of Aristotelian philosophy with Catholic doctrine. Within a few decades of the works of Maimonides being burned under their auspices, the Dominican scholar St Thomas Aquinas (1225–1274) was freely using the Jewish sage, not to mention Muslim philosophers, as authorities in his works of systematic theology. For the Dominicans and Thomas Aquinas, the Globe was never in doubt. Catholic scholars had learnt it, together with the basics of Aristotelian cosmology, from ancient Latin sources. However, this nugget of ancient wisdom took a circuitous route through the furthest reaches of what had been the Roman world before it became widely accepted by early medieval Christians. Somewhat incongruously, the books that cemented the Globe into the Western consciousness were written in what is now a suburb of Newcastle upon Tyne in northeastern England.

17

Europe in the Early Middle Ages: 'Equally round in all directions'

One day in around AD 390, during the season of Lent, St Ambrose of Milan (339–397) stood up to deliver a programme of sermons on the subject of the six days of creation. Bishop Ambrose was a notable thinker in his own right, deeply immersed in the Latin literary tradition built around the works of Cicero. Still, as a clergyman, he had to deliver numerous sermons and didn't have time to write them all from scratch. So, for this occasion, he adapted the previously mentioned homilies on the first chapter of Genesis by his near contemporary, Basil of Caesarea. Like most educated people in the eastern parts of the Roman Empire, Basil had spoken and written in Greek. In the West, the dominant language was Latin, but the social class of the bishops was much the same as it was in the East – they tended to be members of the old nobility. Ambrose was typical. He was serving as a governor in northern Italy when he was installed as bishop by popular proclamation.

In his preaching on Genesis, Ambrose assured his Milanese congregation that they did not need to concern themselves with finer questions of pagan cosmology. It wasn't that he thought there was anything wrong with the subject. Indeed, if philosophers understood the elegant way that the Earth was suspended in the midst of the heavens, so much the better. It gave them an opportunity 'to magnify the excellence of the divine artist and eternal craftsman'. But Ambrose was addressing common people and so downplayed his own formidable erudition, assuring his flock that they didn't require scientific expertise to be good Christians. He was content to warn them against excessive literalism when they were reading the Bible: 'When we read God saying "I have established the pillars of the Earth," we cannot believe that the world was supported actually by columns, but rather that His power props up the substance of the Earth and sustains it.'[1]

Latin-speaking theologians like Ambrose thought the Earth was round, but nevertheless believed the shape of it to be irrelevant. However, a confidant of Emperor Constantine, called Lactantius (*fl. c.* 300–*c.* 330), disagreed. In a lengthy defence of Christianity, he had rebutted many claims taken from the works of pagan philosophers. One of the doctrines he dismissed was the Globe, which he found simply incredible. He was perfectly aware of the evidence and explanations that purported to prove it but mocked them mercilessly. Pagan philosophers, he suspected, 'are philosophizing for fun or else knowingly and wittingly taking up the defence of untruth to practice and display their talents on rubbish'.[2] He promised he could prove the Earth must be flat, but unfortunately he didn't have the space to lay out his argument in full.

Lactantius was an influential author. He wrote with a fine Latin style and was called 'the Christian Cicero'. That said, there is little evidence that his peculiar views on the shape of the Earth attracted much notice, either from his peers or in the later Middle Ages. Today, things are different. Unluckily for his literary reputation, being an unambiguous flat-earther is what Lactantius is best known for.

Augustine of Hippo and Western Christianity

If you wanted to get ahead in late antiquity, mastering the art of rhetoric was one way to do it. The father of Ambrose's most famous protégé had exactly that idea when he scrimped and saved to send his gifted son to school in the hope he'd become a lawyer. The plan didn't work out. The young man, Augustine of Hippo, was baptized as a Christian by Ambrose in AD 386 and shortly afterwards turned his back on worldly fortune.

Augustine became the most influential of all Western theologians, his voluminous Latin works forming the bedrock of Catholic teaching throughout the Middle Ages and beyond. He had been brought up a Christian by his mother but, while away from home for his education, rebelled and started dabbling in various cults. He settled on Manichaeism – the sect founded by the Persian Mani, who had lived about a century earlier.

For Augustine, the appeal of Manichaeism was that it explained the existence of suffering. It taught that the material world was the creation of a wicked deity, so it was hardly a shock to find life was often unpleasant. Nonetheless, as he tells us in his autobiography, *The Confessions*, Augustine found serious problems with Manichaean doctrine. Thanks to the education his father had paid for, he was aware that astronomers could accurately predict the movements of the planets. For Mani, this was blasphemy. He had taught that the planets were divine beings whose role was to shepherd the souls of the righteous to God in heaven. To suggest the paths they took across the sky could be determined by mathematics was an abomination.[3] But Augustine knew astronomers could calculate, many years in advance, 'the day and the hour of the eclipse, and whether it would be total or partial'.[4] Even the most respected of Manichaean preachers could not reconcile their teaching with the success of astronomy. Coincidentally, there were two partial eclipses of the Sun visible from Carthage in 378 and 381.[5] Augustine could not reconcile their predictability with his religious beliefs and suffered a crisis of faith, eventually leading him to fall away from Manichaeism.[6]

In the meantime, Augustine had moved to Milan. There, he met Ambrose and, under the bishop's influence, began to reconsider Christianity. His mother joined him in Italy and arranged his betrothal to a girl of quality, which necessitated abandoning a woman who had been his mistress for fifteen years. In the end, Augustine turned his back on matrimony and did not wed his fiancée. Instead, he was baptized as a Christian and, after returning to Africa, was ordained as a priest four years later. In AD 395, he became Bishop of Hippo, not altogether willingly, and held the see until his death.

Augustine never lost his interest in science, and he praised mathematics as a way to help our minds grasp immaterial concepts.[7] Years later, after he had become a bishop, he worried about his flock falling into the same trap as the Manichaeans, by treating demonstrably false claims about the natural world as religious dogma. In his *Literal Meaning of Genesis*, he warned that pagans knew a fair bit about astronomy: 'Now it is quite disgraceful and disastrous, something to be on one's guard against at all costs, that they should ever hear Christians spouting what they claim our Christian literature has to say on these topics, and talking such nonsense.'[8]

Augustine wrote no less than three commentaries on Genesis in the course of his career, repeatedly making the point that Christians must not allow themselves to become hostages to scientific fortune. He discussed questions such as why the Moon waxes and wanes, the shape of the heavens and whether they rotate. He rarely provided definitive answers. His purpose was not to educate his readers on the facts of astronomy. He just wanted to show that whatever contemporary science said, the Bible was compatible with it.

It's abundantly clear that Augustine knew about Aristotle's world picture of a spherical universe, but he equivocated over whether it was true. The sky might be a lid rather than a sphere, he wrote, and the Moon might emit its own light rather than reflect the rays of the Sun.[9] It didn't matter which was right, because the Bible could be read to conform to either theory. It was written in everyday language, not as a scientific textbook, so Christians should not let themselves be drawn into arguments about whether a particular scientific theory is true or false.

When it came to the shape of the Earth, Augustine kept his counsel. He didn't try proving that the Globe was consistent with the Bible. Nor did he argue that the biblical authors assumed the Earth is flat. The reason for this silence is not clear. Augustine knew pagan philosophers had declared the Earth was spherical, and he knew this was a conclusion that sat uncomfortably with a literal reading of the Bible. He may have decided, quite deliberately, that he just didn't want to go there. It wasn't a hill on which he was willing to die, or even suffer a slight bruise to his considerable ego. That said, modern experts are sure that Augustine accepted the Earth was a sphere. This seems clear from his respect for Greek astronomy coupled with unguarded comments he occasionally made on the subject.[10]

As Augustine lay dying in his home city of Hippo in northern Africa, an army of Vandals was encamped outside. Before long, they would conquer the entire province. The Vandals were one of various barbarian tribes from the east that moved into Roman territory during the fourth and fifth centuries. While the Byzantine Empire managed to fight off or integrate the intruders, the western empire gradually disintegrated under pressure from the invaders. Rome itself was sacked in 410 and again in 455.

The fall of the western Roman Empire took some time but, ultimately, it was total. The city of Rome itself shrank from a metropolis with a population of a million to a town nestled in the ruins of its past grandeur. The economy collapsed as trade across the new frontiers between barbarian kingdoms diminished. Society became more militarized. Kings owed their position to their aptitude for fighting and as leaders of men. At first, the old Roman nobility (Christians trained in rhetoric and philosophy) ran the civil service for their new masters. But soon they were squeezed out of political power and found a new vocation in the Church.

Recognizing that times were changing, a few of the surviving Roman aristocrats sought to preserve the books of the ancient writers and set out a new education syllabus. This was a course of reading intended for Christians looking to become senior clergy rather than secular bureaucrats. The texts prescribed by Cassiodorus (c. 490–c. 585), a grandee turned monk, included Christian and secular authors, but the point of studying the latter was to help illuminate the former. Cassiodorus himself was familiar with Ptolemy and well aware of the Globe but didn't give the matter much attention.[11]

Isidore of Seville

While monks were still copying classical Latin literature, fewer people were reading it. As for Greek books on science and philosophy, these almost completely disappeared in the West as no one had proficiency in the language to read them. Even awareness of the shape of the Earth became distinctly hazy, if it didn't fade away entirely. The commodious works of the long-serving Archbishop of Seville, St Isidore (c. 560–636), are a good example of the resulting muddle.

Spain had been a Roman province for centuries but, by Isidore's time, it was ruled by Visigothic kings whose ancestors were among the barbarians who had invaded the empire. Although they were Christians, the Visigoths were not orthodox believers. Isidore's family set them straight on this point, converting the dynasty to Catholicism. One of the kings, Sisebut (565–621), cemented his place in the bosom of the Church by launching a persecution against Jews when he came to the throne. By then, Isidore had succeeded his older brother to the see of Seville, which he occupied for nearly forty years.

Isidore was the scion of an ancient and noble lineage, which had successfully repositioned itself as a pillar of the Church – his younger brother was a bishop and a saint, his sister a nun. He read widely, seeking to assimilate pagan and Christian knowledge into encyclopaedias that summarized sacred and profane wisdom. His writing contains vast amounts of information, much of it useful but with more than a sprinkling of the misleading and wrong. He richly deserves his modern appellation as patron saint of the Internet.

Isidore's best-known book is a lexicon called the *Etymologies*, which he completed near the end of his life after decades of work. It includes a section on astronomy, where he adopted a typical Aristotelian world picture, except that he described the Earth itself as a disc. In particular, he defines the Latin words *orbis terrae*, which we've seen could mean either a circle or an orb, as meaning 'round like a wheel'.[12] Many modern scholars have thought that when he referred to the circular *orbis terrae*, he was talking about the inhabited world, rather than the Earth itself. Unfortunately, the text of the *Etymologies* is hard to reconcile with this reading and, in recent decades, opinion has moved from giving Isidore the benefit of the doubt and assuming he must have accepted the Globe, to inferring that he probably didn't.[13]

Things don't get any clearer in *On the Nature of Things*, a guide Isidore wrote on natural science. The title echoed the poem by the Epicurean Lucretius and was a conscious effort to sacralize pagan learning.[14] The result is an incoherent mess. For example, Isidore covered astronomy and geography in detail but failed to marry the two subjects into a consistent world picture. As in the *Etymologies*, he accepted Aristotle's view of a spherical universe, with the Earth at its centre and the planets orbiting it. He discussed the five climate zones and explained how these relate to the Arctic and Antarctic Circles, as well as the tropics of Cancer and Capricorn. He understood the mechanism of eclipses and showed how they are caused by the Moon or the Earth blocking the light of the Sun.[15] The Visigoth king Sisebut was interested in eclipses himself as there had been total lunar and solar occultations in AD 611.[16] He was so impressed by Isidore's erudition that he wrote him a poem on the natural causes of these events.[17]

From Isidore's discussion of the celestial spheres, you'd think he accepted the Globe, although he never explicitly said so. But when it came to geography, his descriptions of the world were more consistent with a flat earth. He illustrated the climatic zones as five circles arranged on a disc, not as bands

Isidore's climate wheel from *De natura rerum*, late 8th–early 9th century.

around a sphere. In his schema, the Arctic and Antarctic zones are in contact, as he explicitly noted in the text.[18] This makes no sense if the Earth is a sphere. Other circular diagrams of the Earth in *On the Nature of Things* earned it the name 'The Book of Wheels'. These illustrations could only have reinforced a reader's impression that the Earth is a plate.

Taken as a whole, Isidore's work suggests that, like the Presocratic philosopher Anaxagoras, he placed the disc of the Earth at the centre of a spherical cosmos. Still, the bishop realized he didn't have all the answers. He referred back to the words of Ambrose of Milan: it isn't necessary to enquire too closely into how God arranged the world.[19]

Isidore's infelicities have driven today's academics to despair. He didn't seem willing to be unambiguous one way or the other. In contrast to classical writers from Pliny to Ptolemy, he provided no evidence for the Earth's shape, so his contemporaries, reading his work without any prior knowledge of Aristotelian cosmology, would have been none the wiser. If Isidore was aware that the Earth is a sphere, he made no effort to impart this knowledge to his readers. The theory of the Globe is something that must be taught and Isidore didn't even try to teach it. The most parsimonious explanation for this is that he didn't believe it.

Isidore of Seville was an immensely influential author, whose *Etymologies* covered knowledge of all things in heaven and earth. When books were scarce and expensive, having everything you needed to know in a single volume was an appealing prospect. One of his avid readers was the mysterious author of the *Cosmography of Aethicus Ister*. This book, written in around AD 700, purports to be based on an earlier travelogue by the eponymous Aethicus Ister, although there is no other evidence that such an Aethicus ever existed.

The *Cosmography* describes the world in a way that resembles the Homeric tradition of a flat disc surrounded by a great

Ocean. It slots a Christian version of hell under the ground and says heaven is adhered to the edge of the earth by great hinges in the north and south.[20] The author relied heavily on Isidore's *Etymologies* for his geographical information, but there is no hint he was familiar with the Globe – he certainly didn't learn it from his close reading of Isidore.[21]

There was ambiguity about the Earth's shape among Irish thinkers, too. The author of the *Cosmography of Aethicus Ister* claims to have spent some time in Ireland, perusing the libraries. He wasn't impressed, dismissing his hosts as 'unskilled workers' and 'ignorant teachers'.[22] This is harsh. The scholarly achievements of early medieval Ireland are rightly celebrated. Nonetheless, seventh-century Irish literature contains no clear conclusions about the Earth's shape. Literate people, who would mainly have been monks, probably did come across references to the Globe in classical Latin sources but may not have recognized them, let alone believed them.[23] If they had realized that ancient books were telling them that the Earth is a sphere, the idea would have been so discombobulating that they could hardly have failed to mention it.[24]

The Venerable Bede

Following all this confusion, in the early eighth century, ambivalence about the Earth's form was finally dispelled by the greatest scholar of the era, who lived in a small English kingdom called Northumbria. Universally known as the Venerable Bede (672–735) (he was canonized in 1899), he spent his entire life in the locality of his monastery of Jarrow, on the south bank of the River Tyne. Remarkably, the little abbey church where Bede worshipped many times a day still stands on the outskirts of Newcastle, though little remains of the rest of his monastery.

Bede's writings ranged across an array of subjects. Today he is best known for his *History of the English Church*, from which so much of our knowledge of early medieval England must be gleaned. However, across Europe, the study of the calendar was the foundation of his fame. He consulted Isidore, but when it came to understanding the constitution of the world, he turned to Pliny the Elder's *Natural History* – 'that delightful book', as he called it.[25]

Bede took his description of the Earth from Pliny and, conscious of the risk of misunderstanding, decided to be crystal clear. He wanted his readers to comprehend how the Earth's shape determined matters such as the height of the Sun and the time different stars rose above the horizon. The Earth, he said, 'is not merely circular like a shield, or spread out like a wheel, but resembles more a ball, being equally round in all directions'.[26] The precision of Bede's language, so conspicuously lacking in Isidore, is precisely what we would expect from an author relaying a startling fact. As evidence for the curvature of the Earth, he noted that the times the sun rises and sets depends on where you are. Furthermore, the star Canopus is visible only in the south, from Egypt, and not from Italy or anywhere further north. Bede himself had never seen Canopus, or even been to Italy. Instead, he travelled the world through his books, finding in Pliny's massive compendium all he needed to know.

The reference Bede made to the Earth not being like a wheel might be a direct rebuke of Isidore for muddying the waters. Even as he lay on his deathbed, Bede continued to excerpt Isidore's *On the Nature of Things*. 'I do not wish my students to read falsehoods,' he said, 'or to work at this task in vain after my death.'[27] Bede needn't have worried. His book on the calendar was copied in scriptoria across Europe. It was essential to the acceptance of the Globe that a writer as esteemed as Bede was so unequivocal about it. It also helped that he wrote about it in a book concerned with the date of Easter, a subject close to the hearts of Christian clergy.[28]

In the ninth century, scholars attached to the court of the mighty Frankish emperor Charlemagne (*c.* 747–814), who had claimed the mantle of the Romans, began a cultural revival. They copied classical textbooks by the likes of Macrobius and Martianus Capella, reintegrating ancient learning into the curriculum.[29] Bede's work added a Christian voice to the pagan choir, helping to assuage any doubts about the compatibility of Aristotelian cosmology with the Bible.

Charlemagne himself was a student of astronomy and, like Sisebut the Visigoth, was especially eager to understand eclipses. He wrote to an Irish monk, Dungal of St Denis (*fl. c.* 810–*c.* 828), asking for an explanation of two that occurred in AD 810. Dungal responded with a short treatise outlining the spherical model of the universe and how the Sun, Moon and Earth interact to cause solar and lunar eclipses.[30] Dungal's tract was copied and furnished with illustrations that allowed it to be used as a teaching aid for less august students than the emperor. Thus, despite the obscurantism of Augustine and the bewilderment of Isidore, after the eighth century the Globe became part of the world picture of medieval Christians without much more debate.

The Antipodes

While they had resolved the confusion over the form of the Earth, Christians still had to decide whether there was anyone living on the other side of it. Was there a counterweight continent? Was it inhabited? And, if it was, was everyone who lived there upside down?

The word 'antipodes', meaning 'opposite feet', first appeared in a Latin text written by Cicero, who coined the term to mock the concept that 'there are people opposite to us on the contrary side of the Earth, standing with the soles of their feet turned in the opposite direction to ours.'[31]

Pliny the Elder disagreed that antipodeans were implausible. He noted that the intelligentsia said the Earth was inhabited all round, even though vulgar people wanted to know why dwellers 'down under' didn't fall into space. Pliny pointed out, quite reasonably, that if there was anyone on the other side of the world, they would ask the same question about us.[32] Nonetheless, despite the Greeks having no shortage of legends about fantastic voyages, and not a few explorers who had travelled far to the west and east, nobody claimed to have visited the antipodes. For Pliny, this was unsurprising. As far as he was concerned, intercourse between the northern and southern habitable zones was made impossible by the heat of the tropics that separated them. Even if there were any Australians, no one had met them.

In his *Commentary on the Dream of Scipio*, Macrobius concurred that there were people everywhere on Earth, including in the Southern Hemisphere. He also addressed the perennial concern about antipodeans being upside down. 'Have no fear that they will fall off the Earth into the sky, for nothing can ever fall upwards. If for us, down is the Earth and up is the sky, then for those people as well, up will be what they see above them and there is no danger of them falling upwards.'[33]

In contrast, early Christian authors' condemnation of speculation about antipodeans was unequivocal. Lactantius, who thought the Earth was flat, clearly had no time for the idea that people might live on the other side of it. He blamed the concept of the Globe, which he also regarded as nonsense on stilts, for all this foolishness about upside-down people.[34] As for Augustine, he explained that even if the Earth were a sphere, that still didn't justify a belief in the antipodes. It was quite possible that the entire Southern Hemisphere was ocean. And should there be some distant continent, no one could have survived the long voyage to sail there.[35] Isidore thought it was physically impossible for there to be any antipodeans, dismissing them as the conjecture of poets.[36] Even the Venerable Bede, who was crystal clear that the Earth was round like a ball, said there was simply no evidence for lands on the other side of the world. He couldn't see why anyone would argue these places existed when no one had been there.[37]

Scepticism about the antipodes extended to the top of the western Church. In 748, shortly after Bede's death, the pope heard that a certain Virgil of Salzburg, one of many Irish missionaries in Germany, supported the idea that there were inhabited lands on the other side of the world. His Holiness angrily demanded Virgil's defrocking if it was true that he believed this. We don't know what Virgil really thought about the issue but, whatever it was, his subsequent career didn't seem to be harmed by the controversy. He became a bishop in 766 and a saint in 1233.[38]

Historians used to assume that doubts about the existence of antipodeans were evidence medieval people thought the Earth was flat. However, the debate about the antipodes was still raging long after Western Christians defaulted to acceptance of the Globe. There was also plenty of room for disagreement about the arrangement of the continents and the width of the Ocean. By the end of the Middle Ages, some of these questions would assume momentous importance.

18

High Medieval Views of the World: 'The Earth has the shape of a globe'

Remember the astrologers who calculated the most auspicious date for the foundation of Baghdad? We saw how one of them, al-Fazari, translated an Indian treatise on astronomy into Arabic, thus introducing the theory of the Globe to the court of the Abbasid caliphs. Another was Mashallah, born in Basra to a Jewish family in the early eighth century. He wrote many volumes of practical astrology and lived until after 813, by which time the Globe was firmly established among the circle of astronomers active in Baghdad. As well as astrology, there is a tract on Aristotelian cosmology included in his oeuvre, called the *Book of the Orb*. This contains a full account of the Globe, together with evidence to prove it. Until recently, only a Latin translation of the Arabic original of this book was known. However, in 2011 a philologist from the University of Tokyo announced he had found no less than three copies in Arabic.[1] This was good news for scholarship but rather unfortunate for Mashallah, since it became apparent that he couldn't have written the text after all.

The uncovering of the Arabic version of the *Book of the Orb* shows there is plenty still to be discovered in the world's libraries and archives. Historians call the basements and vaults in which most manuscripts now reside 'the coalface'. There are rich seams to be mined by scholars who have the necessary skills in ancient languages and palaeography (the art of deciphering old handwriting). Who knows what lost works by the likes of Aryabhata or Avicenna might still turn up?

From a review of the Arabic *Book of the Orb*, it looks as if it was written in Egypt, 150 years after Mashallah's death, by a Jewish physician called Dunash ibn Tamim. We've met him briefly because he was one of the earliest Jewish authors to follow in the footsteps of Saadia Gaon in writing about the Globe. His lost guide to Aristotelian cosmology written for the sultan, *The Configuration of the Orb*, is probably none other than the *Book of the Orb* previously attributed to Mashallah.[2] Since Dunash ibn Tamim's book was a useful primer on astronomy, it found its way to Spain via the correspondence networks that linked together Jewish scholars from Baghdad to Toledo. At the time, much of Spain was under Muslim rule and hosted a thriving Jewish population. Dunash's name somehow dropped off the manuscript, which was instead ascribed to the more famous astrologer Mashallah. Then, after another century or so, the book was rediscovered in the libraries of Toledo after the city was occupied by a Christian king.

How Greek and Arabic Learning Reached Europe

In the eleventh century, Islamic Spain was wracked by civil wars and split into a shifting quilt of minor principalities. To the north of the Iberian Peninsula, Christian kingdoms jostled for power, playing off the Muslim princelings against each other, and all the while increasing their territory. In 1085, shrewd diplomacy allowed King Alfonso VI (1040–1109) of León and Castile to take the great city of Toledo almost unopposed. It had once been the capital of Visigothic Spain and now, once again, became an important Christian centre.

Since at least the time of Bede, western Christians had been well aware of the basics of Aristotelian cosmology, including the Globe. But they also knew how far behind the scientific curve they had fallen.[3] The textbooks by Macrobius and Martianus Capella gave them an idea of what they were missing because they mentioned Greek authors like Ptolemy and Euclid, whose works didn't exist in the libraries of Catholic countries. This emphasized the intellectual 'poverty of the Latins' caused by their loss of Greek scholarship after the fall of the western Roman Empire.

The capture of Toledo by a Christian king gave the West a chance to catch up. Scholars from across Europe converged on Spain to translate Arabic books, many themselves translations from Greek unknown to the Latin world for hundreds of years. The libraries of Greek-speaking parts of southern Europe, including Sicily, also came under Catholic control following an invasion by a band of Norman adventurers. Finding the original versions of the celebrated books from antiquity, as well as Arabic commentaries on them, gave scholarship new impetus. Thus, western Europe enjoyed a translation movement, four centuries after Abbasid Baghdad had hosted a similar effort to assimilate Greek philosophy and science.

The outstanding contributor to the task of rendering Arabic into Latin was an Italian cleric called Gerard of Cremona (c. 1114–1187). After learning as much as he could from Latin sources, he travelled to Toledo searching for a copy of Ptolemy's *Almagest*. On arrival, he mastered Arabic and set about translating dozens of books into Latin, which were quickly distributed across Christian Europe. After his death, his students made a list of all his translations.[4] As well as Ptolemy, it includes Aristotle's *On the Heavens*, Avicenna and Archimedes. Among less celebrated works, the catalogue features the *Book of the Orb*, which Gerard must have tackled relatively early in his career because we find it being used in Normandy in the 1140s by one William of Conches (c. 1090–c. 1154).[5]

William was a central figure in an effervescence of new philosophical thinking in the cathedral cities of northern France. After teaching at Chartres, southwest of Paris, he was hired by Geoffrey Plantagenet (1113–1151), the powerful Duke of Normandy, as tutor to his sons. One of the boys would become King Henry II (1133–1189) of England in 1154. William addressed Geoffrey in the introduction to the *Dialogue on Philosophy* that he wrote at the time. He 'set out to write for you and your sons something suitable for scientific studies', going on to praise the duke for encouraging his children to engage with academic subjects rather than playing games all day.[6] To be fair, the boys would have spent most of their time being schooled in the arts of war since the family were, above all, soldiers. This meant that William had to aim his teaching not at would-be scholars but members of a militarized aristocracy.

The *Dialogue on Philosophy* purports to record a conversation between a philosopher and a duke, standing for William and his patron, Geoffrey Plantagenet. The duke asks questions and seeks clarifications, which we can assume echo the kinds of queries William found himself responding to during his long teaching career. One of the issues he had to address was the theory of the Globe. 'I do have some doubts about the shape of the Earth,' the duke admits, before asking the philosopher to explain how we know it is a sphere.[7] Clearly, William couldn't assume his students knew the world is round, so he went back to first principles. As well as the standard arguments transmitted by Bede, he found the newly translated *Book of the Orb* was ideal source material.[8]

The *Book of the Orb* contained several arguments to show the Earth is not flat and buttressed them with diagrams. Plucking one of the examples from the older book, William asked his readers to imagine two cities on a flat world, one in the Far West and the other in the Far East. As shown in a figure, shortly after the Sun rises, it will stand directly above the eastern city. It must then travel across the sky so that it is over the western city just before it sets. In fact, the Sun is everywhere at its highest in the sky at midday, so the Earth cannot be flat.[9] Admittedly, the argument only works if the Sun is quite close to the Earth, but William's main point was that the Globe is counter-intuitive. His readers must not think 'like animals trusting their feelings ahead of reason', who assume the Earth is flat because they cannot sense its rotundity in everyday life. Even high mountains and deep valleys don't detract from the smooth roundness of the world.[10]

William doubtless used his Dialogue as a teaching aid, even though the Plantagenet boys probably never read it. The illustrations and lucid explications meant that it remained a popular work among aristocrats for centuries. The most resplendent surviving copy was created for King Wenceslas IV of Bohemia in 1402.[11] It features illustrations that are so sumptuous they lose some of the clarity that had distinguished the original version. However, the Dialogue was very far from the most common textbook covering astronomy in the Middle Ages. That prize must go to a short pamphlet for university students by an Englishman called John Sacrobosco (d. c. 1256).

The invention of the university was one of the most lasting achievements of the medieval period. Universities organized themselves as corporations responsible for their own internal governance, charging fees to students so they didn't depend on the patronage of fickle rulers. This made them effectively immortal.[12] Indeed, most medieval universities survive to this day and the legal form of a corporate body dominates higher education across the world, not to mention all manner of businesses.

Something like a million students passed through the doors of the universities during the Middle Ages.[13] They arrived at

the age of fourteen, so couldn't be expected to master such difficult authors as Aristotle or Ptolemy without some remedial classes. A genre of short textbooks in simple Latin catered for this market. For astronomy, students depended on Sacrobosco's *The Sphere*. We know next to nothing about the author except that he taught at the University of Paris in the thirteenth century. His textbook was a pithy summary of the universe according to Ptolemy, shorn of the mathematics and rounding off with an explanation of eclipses.

The first section of *The Sphere* describes the Aristotelian cosmos as a giant ball with the Earth a tiny speck in the middle. It goes on to set out several arguments for the Globe. For such a short work, *The Sphere* is reasonably comprehensive, packed full of the information that students needed to know to get their degrees.[14] Postgraduates would encounter Aristotle's *On the Heavens* directly, as well as commentaries on it by scholars such as St Thomas Aquinas. Admittedly, most students never got that far, leaving after a year or two to pursue careers in law or the Church. However, we can be sure that anyone with a modicum of education learnt about the Globe early in their

Giovanni di Paolo, *The Creation of the World and the Expulsion from Paradise*, 1445, tempera and gold on wood.

careers. Nobody seemed terribly bothered that the authors of the Old Testament had never heard of it.

It's one thing to show that the higher aristocracy and anyone who went to university learnt the Earth was a sphere, quite another to ascertain what ordinary people thought. Sacrobosco's *Sphere* and the *Dialogue* of William of Conches, like almost all academic texts before the seventeenth century, were written in Latin. To be 'literate' meant knowing the ancient language rather than, say, English, French or German. The children of the bourgeoisie learnt to read their vernaculars through stories from the Bible and prayers, which obviously didn't teach them about cosmology. Many people couldn't read at all. If we are going to find evidence that awareness of the Globe had spread well beyond the elite, we will need to sweep across a varied range of literature and other kinds of art. For instance, when the Sienese painter Giovanni di Paolo (d. 1482) wanted to illustrate God creating the universe, he could portray this according to the Aristotelian world picture, knowing it would be recognizable.

Common Knowledge?

Around the start of the twelfth century, a loose group of singers began to tour southern France and the surrounding territories. They became known as troubadours. These artistes might be commoners or members of the nobility: the only qualification to join their ranks was a voice to hold a crowd entranced. Troubadours sang about knights and damsels, setting their tales in the eras of Charlemagne, King Arthur or the Trojan War. According to surviving poems, they occasionally alluded to the theory of the Globe in passing, while they regaled their listeners with the deeds of heroes from the past. For example, in one of the French romances, Alexander the Great received a ball as a gift, which he took to represent the world he would shortly conquer.[15]

However, like a good TV channel, the troubadours didn't limit their repertoire to drama and romance; they also did documentaries. There is a genre of Old French didactic poems dating to the thirteenth century that gave listeners a précis of what was known about the world. These referred to the Earth being 'round like an apple', leaving listeners in no doubt that it was a sphere.[16] Unfortunately, we know little about the environment in which these poems were performed and who was listening to them. We can be confident, though, that they had a larger footprint than university textbooks, much as trade publications today sell more copies than academic monographs.

Dante Alighieri (1265–1321) was a poet more celebrated than any of the troubadours. He was born in Florence but was exiled in 1301 as a result of the city's fractious politics. By writing in the dialect of Tuscany, which would eventually morph into modern Italian, he was able to reach an audience among the merchants in Italian city states. His most famous poem, the *Divine Comedy*, follows Dante on a journey through hell, purgatory and heaven. He artfully combined the cosmology of Aristotle with a Christian vision of the afterlife, depicting the universe as a hierarchy rising from the depths of hell at its centre to the beatific vision of God beyond its furthest edge.

The first part of the *Divine Comedy*, *Inferno*, commences on the surface of the Earth, where Dante finds himself lost in a wood. He is joined by the pagan poet Virgil, who takes him on a tour downwards through the nine circles of hell. At hell's very centre, they find Satan, frozen up to his waist in a lake of ice. Emerging on the other side of the world, Dante stands beneath the great mountain of purgatory. In Catholic theology, this is where the souls of the blessed dead are purged of their sins before they can ascend to heaven. Catholics teach that purgatory is a place of punishment but also of joy as the suffering souls know they are destined to escape it in the fullness of time. During the second part of the *Divine Comedy*, Dante ascends through the mountain's terraces on his way to the sphere of the Moon. This is the threshold of heaven.

Aristotle had explained that, unlike the Earth, the celestial objects above the Moon were incorruptible, which, to Christian minds, made them a suitable locale for heaven. Dante arranged them into nine spheres, occupied by the seven planets, then the fixed stars and, finally, the outermost sphere that bounded the universe. Beyond, he placed the abode of God, just as Aristotle had done. *Paradise*, the third part of the *Divine Comedy*, recounts Dante's ascent through the spheres in the company of Beatrice, a virtuous Christian lady and his lost love.

It's pretty obvious that Dante drew heavily upon Christian theology and ancient Greek cosmology when he created his vision of the eternity. However, the closest parallel to his world picture that we've encountered so far in this book was created in China at about the same time as he was writing. This is the Manichaean *Diagram of the Universe*, which we admired earlier. There are obvious differences between the *Divine Comedy* and Manichaean doctrine. For a start, they disagreed about the shape of the universe. Dante follows Western medieval consensus by building the universe from spheres. The Chinese Manichaeans imagined a stack of worlds, piled on top of each other vertically. Get beyond that basic difference, though, and similarities emerge. Both world pictures portray a hierarchy of heavens and hells, with multiple tiers above and below the surface of the Earth. Both show a mountain crossing the void between the heavens and earth. They share a strong moral sense of direction – up towards the virtue and down into the midst of wickedness. Each assumes that the universe was fashioned along ethical lines, with the ideals of good and of evil anchoring its top and the bottom.

There's no need to postulate any direct influence on Dante by Mani.[17] After all, the stratified universe goes back to ancient Babylon. We've found it in the Puranas, the Koran and the Talmud. In all these cases, the Earth was the filling sandwiched between heavens and hells. Dante reconfigured the tiered universe to show that theological meaning could, just as well, be contained within a spherical cosmos, with the Earth a ball at its centre. 'Down' pointed to the centre and 'up' towards the rim.

It's true that sacred texts could inhibit the adoption of the Globe by their inspired authority, but there was rarely anything religiously incommensurable between a traditional and spherical world picture. Dante showed that the theological concepts previously illustrated with a tiered universe could just as well be communicated by one shaped like an onion.

Towards the end of the Middle Ages, prose writers also had occasion to mention the Globe. One instance was a book about geography called *The Travels of Sir John Mandeville*, which circulated in French and English versions. Dating from the mid-fourteenth century, it tells of a trip by its author from his home in northwestern Europe, across Asia to the Indies. Despite its manifold inaccuracies and frequent infelicities, the *Travels* was extremely popular.

We know nothing about the writer of this fantastical travel guide, except that he definitely wasn't called Sir John Mandeville, as no such person existed. In truth, the author had never ventured far from home and garbled what he had read about more distant countries. Still, he evinced no doubt that the Earth was a sphere. In an aside, while discussing the islands of Southeast Asia, he explained that, because the world is round, the pole star cannot be seen from the Southern Hemisphere. Instead, people looked to another star over the South Pole to navigate. This star wasn't visible from northern latitudes but could be spotted from Libya.[18] In reality this 'Antarctic pole star' doesn't exist and may be a misconstrued reference to Canopus. It's clear that the author of the *Travels* cheerfully accepted that the whole world is inhabited, mocking those of 'limited understanding' who worried that antipodeans might fall off into space.[19]

The *Travels* relays a couple of estimates of the Earth's circumference, ranging between 32,000 and 48,000 kilometres (20,000 and 30,000 mi.), meaning it was quite possible to sail all the way around. It features a yarn about a chap who almost managed to complete the circumnavigation but couldn't find passage for the last leg of his journey and had to turn back the way he came. Nonetheless, the author of the *Travels* had trouble understanding all the implications of the Globe. Even while he was explaining the Earth was spherical, he imagined Jerusalem

Coronation of Harold II, detail from the Bayeux Tapestry, late 11th century.

was not only at its centre but its highest point. To reach the holy city from Europe, he said, you had to climb uphill, while the onward journey to India was downwards.[20]

Evidence of widespread knowledge of the Globe in the Middle Ages isn't confined to the written word. While most people couldn't read, they could still look at pictures and objects. Take the coronation orb, which has been part of the regalia presented to kings and emperors from late antiquity. The orb represents the Earth, while the cross on top symbolizes that the secular power of the monarch was subject to the will of God. Presumably, if people in the Middle Ages had thought the Earth was flat, they would have presented their rulers with a dinner plate instead.[21] There's a scene in the Bayeux Tapestry of Harold's coronation as king of England at Westminster Abbey in 1066, showing him holding the orb. The tradition continues to the present day – King Charles III was handed a similar accoutrement when he was crowned on the same spot in 2023.

A Round World

The multiple strands of evidence that show medieval people knew the shape of the Earth does not mean they saw their world in the same way that we do ours. By far their most common visualization was the circle of continents ultimately derived from Homer. Isidore of Seville had designed an elegant representation of this scheme called a T-O map. The 'O' is formed by the circle of the Ocean, orientated with the east towards the top. This is cleaved by two lines, one running across the middle, north to south, and the other branching down from its centre to the west. Together, these lines form the 'T', splitting the circular landmass into three. The upper semicircle represents Asia. Below, Europe and Africa are separated by the stem of the 'T', which denotes the Mediterranean Sea. Its crossbar stands for the rivers Nile and Don. One consequence of this arrangement is that Jerusalem is usually located near the centre of the world.[22]

Isidore's influence ensured that his T-O maps enjoyed an afterlife, separately from his books, as a principal model of medieval geography. Natural philosophers later devised an argument based on Aristotelian cosmology to explain why the inhabited world had to be circular. We've seen that Aristotle had said that the Earth settled at the centre of the universe because heavy matter naturally sank there. Water, the next heaviest element, constituted a shell around the Earth. Then came air, giving us the atmosphere. Finally, just below the Moon, there was an invisible layer of fire, the most rarefied of the elements. This implied that the shells of water, air and fire had to be larger than the Earth they surrounded. Characteristically, philosophers

T-O map from
Isidore of Seville,
Liber etimologiarum
(1472).

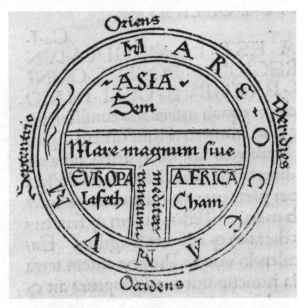

were guilty of overthinking this issue. They worried the continents would be completely submerged beneath an endless
expanse of ocean.[23] One solution, which became dominant after
1400, was to assume the two spheres weren't concentric. The
Earth poked out from the Ocean on one side, which meant the
area of dry land would be round, just like a T-O map showed.
Incidentally, this theory also made it impossible for there to be
any continents on the other side of the world. No one seemed
perturbed that Aristotle himself had ridiculed the idea that the
inhabited world was a circle.[24]

The simple T-O map was an easy way to represent the inhabited part of the Earth. It also gave its basic form to the more
detailed *mappa mundi*, or 'cloth of the world' in Latin. The largest
surviving example hangs in the library of Hereford Cathedral.
It shows the Earth as a flat membrane attached to the ring of
the universe by clips, as if it is a giant trampoline. Although you
couldn't describe it as geographically accurate, the Hereford map
is full of colour and detail that illuminates the Christian vision

The Hereford Mappa Mundi, c. 1300, vellum.

of the world. It isn't just a map of places; it ranges across time, showing important past events. As is typical, Jerusalem is in the middle, where Jesus hangs from the cross. Noah's Ark balances on a mountain range in eastern Anatolia, while Egypt hosts the camp of Alexander the Great. Everything is carefully labelled, so a monk, looking at it in 1300, shortly after it was created, would have been able to locate himself on the map at Hereford. From there, he could have seen where he stood in relation to the entirety of both sacred and secular history.[25]

A hundred and fifty years later, European knowledge of geography was more comprehensive but could still be encapsulated into a circular map. The oligarchs of Venice, in particular, had a good idea about where the spices that they acquired in the Middle East had come from. Their most impressive map, created by a certain Fra Mauro in around 1450, is the apotheosis of medieval cartography. Orientated with south at the top, Europe is easily recognizable in the bottom right quarter. Asia and Africa, on the other hand, are strangely distorted to fit into the round frame. Fra Mauro himself was a monk from the island of Murano in the Venetian lagoon. While he did not maintain that the continents formed a perfect circle, he filled in the Ocean at the edge of the inhabited world with islands to trace the overall round shape. The map shows no sign that the Earth is spherical, but diagrams in the corners of its frame illustrate Aristotelian cosmology, as well as the Garden of Eden.

T-O maps and *mappae mundi* lasted until the sixteenth century, when the European encounter with the Americas finally rendered them obsolete. However, medieval people didn't always portray the continents as a circle. They knew enough about topography to realize this might not be the true configuration of the Earth's landmasses. Travel was more common than we might suppose, whether for trade, soldiering or diplomacy. A few hardy souls had made it to China and Mongolia. There are even records of individuals from the Far East travelling to Europe, one of whom enjoyed an audience with the pope.[26] This meant that Asia was understood to be vastly longer than Europe, and some medieval world maps reflected this. They could be oval, almond-shaped or rectangular.[27]

One lineage of maps, originating in northern Spain, included a fourth continent shaped like a crescent tucked down beneath Africa. This particular figment derived from an off-the-cuff remark made by Isidore of Seville.[28] Although he rejected the antipodes as physically impossible, he did concede that there might be a fourth continent to the south, on the same side of the Earth as us, very hot and separated from the inhabited world by impassable seas. He thought this isolated land might account for the persistent rumour that there were antipodeans.[29] Explorers would scour the southern seas until the eighteenth century, searching for *Terra australis*, the mysterious continent believed to await discovery.

The modern picture of the Earth's surface, with the familiar pattern of continents, began to form in the fifteenth century with the recovery of the *Geography* of Ptolemy. However, the voyages of exploration, starting in the fifteenth century, showed that the surface looked very different from people's expectations.

19

Columbus and Copernicus: 'New worlds will be found'

At the Louvre in Paris hangs a painting by Emanuel Gottlieb Leutze (1816–1868) called *Christopher Columbus before the High Council of Salamanca*. Leutze was the son of German immigrants to the United States but returned to Europe to train as an artist. He continued to depict American themes, including a whole series of pictures on the career of Columbus. The scene in his *High Council of Salamanca* comes from a biography of the explorer by Washington Irving (1783–1859) called the *History of the Life and Voyages of Christopher Columbus* (1828). In the painting, Columbus stands to the left of a table piled with evidence for his contention that he could sail west across the Atlantic Ocean to India. He faces a junta of churchmen, among them a monk clutching his Bible defensively and a cardinal reading from an enormous book.

Washington Irving was a celebrated American author of romantic short stories. While visiting Spain, he was given access to the country's official archives, which he mined for the material that went into his Columbus biography. It was a commercial success, going through over a hundred editions. And it cemented into popular consciousness the trope that Columbus had proved the Earth is round. In the vivid episode that inspired Leutze's painting, the explorer has to face down opposition from a committee in Salamanca, convened to decide whether the Spanish crown should finance his voyage. In Irving's telling, the clerics

Emanuel Gottlieb Leutze, *Christopher Columbus before the High Council of Salamanca*, 1841, oil on canvas.

were not very consistent in their objections to Columbus. Initially, they quoted Lactantius at him, on the impossibility that the Earth is globular, then:

> To his simplest proposition, the spherical form of the Earth, were opposed the figurative texts of scripture. They observed that in the Psalms the heavens were said to be extended like a hide, that is according to commentators, the curtain or covering of a tent ... and that St Paul in his Letter to the Hebrews compares the heavens to a tabernacle or tent, extended over the Earth, which they thence inferred must be flat.[1]

As this failed to dissuade Columbus, the experts pivoted to concerns about the antipodes, which obviously had nothing to do with his scheme. Irving's story was embellished by later writers, who were scrounging around for evidence of a great conflict between science and Christianity. Columbus found himself transformed from a militant Catholic, seeking to claim new lands for the faith, into a rebel against clerical obscurantism.

Born in Genoa in around 1450, Columbus had spent much of his life at sea, honing his navigational skills with voyages to West Africa, the British Isles and perhaps even Iceland. During the years of trying to drum up funding for his expedition across the Atlantic, he'd been rebuffed in Italy and Portugal. He tried Spain in 1486, where the government-appointed committee at Salamanca rejected his plans. The question of the Earth's shape was no part of the opposition — everyone there knew it was a sphere. And they accepted the theoretical possibility of sailing west to the Indies. The story of a circumnavigation in *The Travels of Sir John Mandeville* might have been a fable, but it was just about plausible.

So why, exactly, were the experts sceptical about Columbus's plan? The trouble was, no one in Europe knew America existed, so they had to assume it would be necessary to sail the entire distance from Spain to Japan in one go. That's about 19,000 kilometres (12,000 mi.) – well beyond the capabilities of a fifteenth-century ship. For his scheme to be practicable, Columbus needed to show the voyage would actually be much shorter. And for that to be the case, he had to be very wrong about the Earth's circumference and the size of its continents. Luckily for him, on this matter, the geographical writings of Ptolemy of Alexandria were in error too.

Columbus and the Size of the Earth

In the fifteenth century, Italian humanists began searching for copies of ancient literature to bring them back into circulation. One of the rediscoveries was *On the Nature of Things* by Lucretius, found in Germany. Another was *Geography* by Ptolemy of Alexandria. Unlike the *Almagest*, it had almost been forgotten, even in the Byzantine Empire, until about 1300, when a copy turned up in the library of a Greek monastery.[2] A century later, in 1406, the text travelled to Florence where it was translated into Latin. As we've seen, most of Ptolemy's *Geography* consists of the latitudes and longitudes of about 8,000 locations. However, as far as its medieval readers were concerned, the conceptual novelty of the work was the instructions it gave on how to project the rounded surface of the Earth onto a flat piece of paper. The resulting world map looked nothing like a *mappa mundi*. Instead, it showed the length of Eurasia bending around the curvature of the Earth. Looking at this map, which was printed thousands of times over the following century, it was possible to visualize how the continents related to the Globe.

One infelicity of Ptolemy's *Geography* was his figure for the circumference of the Earth. As we've seen, during the Middle Ages, the received figure from ancient Greece was 252,000 stades (roughly 50,000 kilometres), as calculated by Eratosthenes. While rather too large, this isn't a bad approximation. However, Ptolemy had erroneously preferred an estimate of just 180,000 stades (33,000 kilometres).[3] This would reduce the size of the Earth by 30 per cent and so cut the westward distance between Europe and Asia to about 13,000 kilometres (8,000 mi.). Columbus found this intriguing, but it was still too far for a tiny caravel to sail non-stop. Luckily, he could avail himself of a further mistake, this one made by an Italian mathematician called Paolo Toscanelli (1397–1482). He wrote

to the Portuguese crown in 1474, while the king was considering whether to sponsor Columbus, inserting a map that showed
Asia extending 8,000 kilometres (5,000 mi.) further east than
it does in reality.[4] Columbus copied the letter and convinced
himself that the distance he had to sail was a manageable 5,000
kilometres (3,000 mi.) or less.[5]

The experts at Salamanca disagreed. They saw no reason to
either reject Eratosthenes' figure for the Earth's circumference,
or accept Toscanelli's conjecture about the length of Asia. As
shown on the Venetian map of Fra Mauro, they held that the
inhabited world was circular and protruded from the sphere
of water. That meant the rest of the world had to be covered in
ocean and the sailing distance from Spain, across the Atlantic,
to the Indies necessarily had to be over half the distance around
the whole globe.[6]

It's true that the committee was being slightly old-fashioned.
By Columbus's time, working cartographers had already moved
on from portraying the continents as a circle of land protruding
from one side of an orb of water. They took their inspiration
from Ptolemy, who thought the Earth was a solid sphere. The
oceans lay in divots on its surface so that it was quite possible for
dry land to be present anywhere in the world. The remarkable
survival of a terrestrial globe, completed in 1492, shows no sign
of the circular landmass of medieval tradition.[7] The globe, called
the *Erdapfel* or 'Earth Apple', was commissioned by a well-
travelled merchant called Martin Behaim (1459–1507) and is
now housed in the German National Museum, Nuremberg. It
was based on Ptolemy's projection, with some adjustments made
to parts of the world where his knowledge was deficient. There
are still some errors: Scandinavia does not connect to the mainland in the north, while the English city of Lincoln has been
relocated to Scotland. The Americas are a conspicuous omission.
Although Christopher Columbus left on his first voyage while

Martin Behaim, *Erdapfel*, c. 1492–4,
parchment, paper, iron and brass.

the *Erdapfel* was being manufactured, he didn't return until the
following year. Still, even if the news about America had arrived
in time, it is hard to see where the new continent could have
been inserted. The Atlantic Ocean is roughly the right size but,
as Columbus had insisted, Asia rather than America is located
on the western side. The *Erdapfel* placed Japan a bit south of
where today we would expect to find Florida.

Behaim was employed as a consultant by the Portuguese
crown, so the *Erdapfel* could reflect, at least in part, the

discoveries of Iberian explorers. The Portuguese might have spurned Columbus, but they had their own programme of reconnaissance. They had spent decades probing south around the shores of western Africa, searching for gold and, ultimately, a sea route to India. In 1488 Bartolomeu Dias (*c.* 1450–1500) doubled the Cape of Good Hope, demonstrating that, contra Ptolemy, it was possible to sail into the Indian Ocean. Columbus himself was present when Dias reported back to the king of Portugal.[8] The trick was to steer deep into the Atlantic to catch the winds that blew towards the Cape, rather than trying to creep down the African coast. A few years later, an expedition carrying out this manoeuvre accidentally discovered Brazil.

The cosmographical implications of the Portuguese voyages were profound. Recall that many ancient writers had maintained that the tropics were too hot to traverse. Portugal's push across the equator was a challenge to this conventional wisdom. It turned out that while the region around the equator was uncomfortably warm, Europeans could survive there. And if the ancients were wrong about that, they might be ignorant of other aspects of geography.

Geopolitics was a major motivation for exploring. Portugal and Spain felt isolated on the western edge of Europe, unable to profit from the rich trade in spices that flowed through Egypt and on to Venice. Columbus had always maintained that a primary benefit of a western route would be direct access to the source of spices in the Far East, undercutting Venetian and Arab merchants. There were martial aspirations at play as well. The Iberian nobility delighted in calling themselves crusaders against the infidel. The Spanish were in the process of conquering the last Muslim redoubt of Granada and finally expelling the Moors from Andalusia. The Portuguese had taken the fight against Islam to North Africa, where they had seized the rich entrepôt of Ceuta. However, at the other end of the Mediterranean,

Islam was in the ascendant. The Ottoman Empire had captured Constantinople in 1453 and was now encroaching far into the Balkans. A sea route to India could outflank the Turks, especially if it enabled Christian Europe to link up with the mythical kingdom ruled by Prester John, located somewhere in Africa or Asia.

With economic, military and religious factors endorsing the case for an alternative route to the Indies, the Spanish monarchs, Ferdinand (1452–1516) and Isabella (1451–1504), overruled their advisers and decided to fund an Atlantic expedition. So it was that, in 1492, Columbus took command of a fleet of three ships and headed into the fathomless ocean. Given his calculations, he wasn't surprised when, three weeks later, he bumped into the islands of the Caribbean almost where he expected to find East Asia. Until his dying day, he believed he'd sailed to India. We still call the islands he discovered the West Indies. Despite his series of mistakes about geography, Columbus's achievement in introducing the Americas to Europe was considerable and its consequences profound. He didn't 'discover' America. It had plenty of inhabitants already. Nor was he the first to navigate the Atlantic. The Vikings had crossed via Greenland and briefly settled in Canada a few centuries earlier. However, Columbus and the other European explorers did link distant countries to each other for the first time. They unified the map of the world.

Once it became clear that Columbus had stumbled upon a continent unknown to Europeans, Spain had to abandon hopes that Japan and China lay just beyond. This was an embarrassment to clerical types, who had to explain how America had escaped notice by the authors of the Bible. As an early sixteenth-century writer chuckled, 'Not only has this navigation confounded many statements by former writers about terrestrial matters, but it has also given some anxiety to interpreters of the Holy Scriptures.'[9]

Classical authorities didn't fare much better. One modern revelled in their obsolescence, crowing, 'If I had Ptolemy, Strabo, Pliny or Solinus here, I would put them to shame and confusion.'[10] Thankfully, the writings of the Romans did seem to include some oblique references to America. It looked like Emperor Nero's Stoic minister Seneca had predicted Columbus would discover America in his play *Medea*:

There will come an epoch late in time, when Ocean will loosen the bonds of the world and the Earth lie open in its vastness, when Tethys [the Ocean's wife] will disclose new worlds and Thule [Iceland] not be the farthest of lands.[11]

One thing Columbus didn't do was 'prove the Earth is round'. Indeed, his voyages didn't even produce new evidence for the Globe. Sailing 4,800 kilometres (3,000 mi.) west to America couldn't demonstrate anything that walking the same distance east to Central Asia had not already done. About the only European who didn't think the Earth was spherical was none other than Christopher Columbus himself. On his third voyage, he made such a mess of his observations of the pole star that he became convinced the world had gone pear-shaped. As he wrote in his account of the expedition, the Earth is not a sphere 'but the shape of a pear which is round everywhere except at the stalk, where it juts out a long way, or it is like a round ball on part of which is something like a woman's nipple'. He seemed to think that the behaviour of the pole star was caused by him sailing up this protuberance. He was quick to add that this didn't mean Ptolemy was wrong: the eastern hemisphere was indeed perfectly round, as the Alexandrian had said. It was only the western hemisphere that projected outwards.[12]

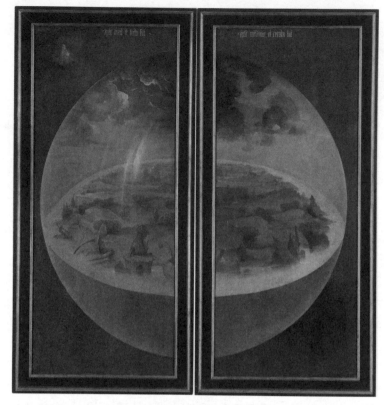

Hieronymus Bosch, *The Garden of Earthly Delights* (exterior panels),
1490–1510, oil on oak triptych.

Nonetheless, even in Columbus's time, there were still those
who found that it suited their purpose to portray the Earth as
a disc. There's a vivid illustration of this on the front panels of
the *Garden of Earthly Delights*, a triptych by the Dutch master
Hieronymus Bosch (c. 1450–1516). The celebrated painting once
formed part of the collection of the Spanish monarchy, and you
can now find it in the Prado Museum in Madrid. You'll be hard-
pressed to get a close look at the depiction of the garden itself,
thanks to an inevitable crowd of tourists. However, walk around
the back, and you can see the panels that are on the front of the
triptych's doors. When closed, they comprise a single image of

a spherical orb, half-filled with water. The waters encircle an uninhabited but fecund continent, which represents the third day of creation in Genesis, when dry land and vegetation first appeared. Unlike Giovanni di Paolo, Bosch took his image from the Bible without any effort to reconcile it with the Aristotelian world picture. He would have known perfectly well that the Earth is a sphere but was still comfortable that an idealized image could show it as a disc.

Around the World

The success of Columbus drove the Portuguese to redouble their efforts to find an eastern route around Africa. In 1497, the brutal and determined navigator Vasco da Gama (*c.* 1460–1524) made it past the Cape of Good Hope and reached the Malabar Coast of India. By most measures, the expedition was a disaster. Vasco da Gama lost the majority of his crew, antagonized the locals almost everywhere he went and barely limped back home. However, the mere fact he succeeded was encouragement enough for the Portuguese to send bigger fleets to the Indian Ocean, where their technological edge and violent tenacity carved out a maritime domain that allowed them briefly to join Europe's great powers.

Then, in 1522, 2,000 years after Aristotle said it was possible, east and west were finally united when someone circumnavigated the Globe. The achievement is usually attributed to Ferdinand Magellan (1480–1521), a Portuguese veteran of the India trade now employed by the Spanish crown. However, he was killed after provoking a fight with a group of Indigenous people in the Philippines, and we can't say for certain which member of his crew was the first person to travel around the world.

Magellan's squadron of five ships had set off from the port of Sanlúcar de Barrameda in southwestern Spain in September 1519 to accomplish the mission of Christopher Columbus and find

a western route to the Spice Islands of Asia. Magellan successfully crept through the straits at the tip of South America into
the open sea beyond. He named these waters the Pacific Ocean,
since they seemed so calm after the storms that battered his fleet
as it fought through the passage from the Atlantic. Beyond the
straits, he set off northwards, searching for winds and currents
to take him west over the Pacific. Sailing across this vast body
of water was an unparalleled feat, which Magellan would never
have attempted had he known how big it was. In the event, it
took three wretched months to make land, having somehow
managed to avoid all the islands that dotted the western Pacific.

After Magellan's death in the Philippines, the remnants of
his crew limped around Africa, eventually reaching Spain in a
single ship three years after they left. Fewer than twenty of the
original complement of 270 made it home. Depending on how
far they had travelled east previously, one or more of these sailors was the first to achieve a circuit of the world. Or perhaps it
was Enrique, a slave originally from Sumatra, whom Magellan
took with him as an interpreter. Following his master's death,
Enrique stayed ashore in the Philippines, but it is not known
whether he returned to his home, thus completing a circumnavigation that ended in the Far East rather than Spain. The
upshot of Magellan's voyage was to show there was no viable
route across the Pacific to East Asia. It wasn't until 1580 that
another expedition successfully completed a circumnavigation,
when Sir Francis Drake (c. 1540–1596) arrived back in England.

The Earth Moves

The voyages across the oceans and around the coast of Africa
had rendered Ptolemy's geography obsolete. It would take a little
longer to overthrow his astronomy as well. Few people ventured
far into the *Almagest* or mastered the calculations required to

predict the movements of the planets. One scholar who did do the hard mathematical yards was Nicolaus Copernicus (1473–1543), a Polish canon who'd been educated in Italy and immersed himself in Ptolemy's complex geometrical models and the Arabic enhancements to his work. In 1543, after many years of rumination, he published *On the Revolutions of the Heavenly Spheres*. This book rejected the ancient axiom that the Earth was stationary at the centre of the universe and proposed that it orbited the Sun. We won't be discussing how Copernicus's idea contributed to the supplanting of Aristotelian cosmology, but it is worth looking at what he had to say about the Globe.[13]

It might seem odd that Copernicus felt he needed to write anything about the shape of our planet. After all, the Globe would have been taken for granted by all his readers. There are a few reasons why he did discuss the topic. The first was his choice to consciously echo the *Almagest*, organizing his book in the same way. This meant starting at first principles with the shape of both the universe and Earth. Copernicus's short chapter on the evidence for the Globe is terse, mentioning the age-old exemplars that Canopus is visible from Egypt and not Italy, and that you can see further from the top of a ship's mast than you can from the deck. Later, he noted Aristotle's observation that the shadow of the Earth on the Moon during a lunar eclipse is curved.[14]

The second reason for bringing up the Globe was that Copernicus wanted to show that opinions about scientific matters change, and that even illustrious clerical writers can get things wrong. 'For it is not unknown', he wrote in his preface, 'that Lactantius, otherwise a distinguished writer but hardly a mathematician, speaks in an utterly childish fashion concerning the shape of the Earth, when he laughs at those who have affirmed that the Earth has the shape of a globe.' This was too much for the Inquisition censors. In 1616, the Catholic

Church rejected heliocentrism as foolish and absurd, not to mention contrary to the Bible. The Congregation for the Index of Forbidden Books reviewed *On the Revolutions of the Heavenly Spheres*, downgrading Copernicus's theory to a mere hypothesis. The tweaks that the censors made to the text were slight, but the reference to Lactantius was struck out.[15] It wasn't that the inquisitors thought Lactantius had a point when he said the Earth was flat. They just didn't want Copernicus poking fun at an esteemed Christian Father and casting doubt over whether the Church knew what it was talking about.

Copernicus's contrivance that the Earth was a planet was impossible to reconcile with the Aristotelian world picture. Aristotle had proved the Earth had to be spherical because it was the coagulation of solid matter that had collected from all sides in the middle of the cosmos. He also implied that the elemental spheres of earth and water should be distinct because they were striving towards the centre of the universe by differing degrees. In making the Earth a satellite of the Sun, Copernicus was invalidating Aristotle's explanation for why it was a ball in the first place. Instead, he said that God required that matter form itself into globes, wherever they happened to be.[16] He couldn't resist a further dig at the Aristotelians for their discredited view that the spheres of earth and water were separate. Both, he insisted, have the same centre of gravity. As Ptolemy had said, the oceans formed from water filling depressions in the Earth, leaving the higher ground as dry land.[17] That meant there was the possibility of undiscovered countries all over the world.

Through the sixteenth century, Europeans continued to explore, using their ships and weapons to dominate many of the people they encountered. As well as military and mercantile opportunities, they sought converts to Christianity. Among the most active of missionary orders was the Society of Jesus, better known as the Jesuits. They followed in the wake of the Portuguese across the Indian Ocean, spreading their faith (with varying degrees of success) across India and Japan, and to the largest and most populous nation of the era, China.

20

China: 'The heavens are round and the Earth is square'

I t was a big job, being emperor of China. He didn't just have
to run the country. The universe itself depended on him. He
had a sizeable support staff of bureaucrats and eunuchs, but the
buck stopped at the dragon throne. Dong Zhongshu (*c.* 175–
c. 105 BC), an adviser to Emperor Wu (156–87 BC), wanted to
ensure his master understood the magnitude of his responsi-
bilities. He explained that the emperor was the link between
Heaven and Earth. If he ruled well, Heaven would support him
by bestowing a clement climate and acquiescent population on
China. However, if he violated the celestial order, for example
by issuing judicial punishments that did not fit the crime, the
emperor could expect warnings in the form of eclipses, droughts
and storms. If he didn't heed the omens, Heaven might with-
draw its mandate and the emperor's dynasty would fall.[1] Because
they were an early warning system of empyrean displeasure,
signs from the sky were of pressing importance, just as they
had been to the Babylonians. From early times, the emperors
maintained an astronomical bureau with the job of interpreting messages in the stars and giving rulers fair notice that their
mandate was fraying.

Chinese chroniclers framed history with the ascent and
decline of dynasties. When a new imperial family took the
throne, it was like a youth that needed to be taught the ways of
rulership, before reaching maturity when the empire was stable

and at peace. But inevitably, dynasties, just like people, grew old and decadent. At that point, a new vibrant lineage would burst forth and overthrow the old, starting their own dynasty afresh. In practice, however, the mandate of Heaven was always awarded retrospectively. It wasn't until a pretender had firmly planted his bottom on the throne that he ceased to be a wicked rebel and became the righteous instrument of destiny. Only then could he call himself the Son of Heaven.

In 221 BC, China was united under the Kingdom of Qin (pronounced 'Chin', from which we get the word 'China').[2] Later historians would look back at the Warring States period before the Qin unification as an epoch of chaos and suffering, but it was also the time when many of the foundations of Chinese culture were laid. In particular, the ideologies of Confucianism and Taoism coalesced and began to enjoy enormous influence.

After the death of the first emperor in 210 BC, his realm threatened to fragment and return China to mayhem. However, one of the generals vying to replace him was able to reassert control in short order and founded the parvenu Han dynasty, which lasted until AD 220. The moral vision of Confucius (551–479 BC), centred on filial loyalty and ritual, undergirded the political philosophy that Dong Zhongshu urged upon Wu. Together, they enshrined Confucianism as the national ethic of China.

The Chinese Universe

According to the *Book of Documents*, one of the four ancient Confucian Classics, the science of astronomy began with an order from the legendary Emperor Yao, who ruled China in about 2300 BC. He sent out four men – one each to the east, south, west and north – to determine the dates of the solstices and equinoxes by observing the heavens.[3] Thus Emperor Yao instituted the first calendar by which all activity in his realm could be directed.

The emperor had to maintain an accurate calendar so that the rites necessary to keep the universe in balance were performed at the designated moments. The consequences of getting this wrong could be catastrophic. In AD 175, government officials reported that calendrical miscalculations had resulted in 'evil folk rebelling and thieving . . . and robbers and bandits making endless trouble'. They demanded repercussions against the errant mathematicians responsible, urging that they 'should receive heavy punishment for empty deceptions'.[4]

Maintaining the calendar wasn't easy. China used a lunisolar system that required frequent adjustments because the periods of the Sun and Moon are incommensurable. That means it is impossible to fit a whole number of lunar months into a whole number of solar years. According to some counts, the astronomical bureau got through almost a hundred calendrical systems in its 2,000-year history.[5] An error of just a day was disastrous if it meant that a potential eclipse was missed or an important ritual was celebrated on an inauspicious occasion. In Europe, the solar calendar desynchronized from the seasons by eight days before the papacy introduced the Gregorian Reform to put it right in 1582. The emperors of China could never have tolerated such an egregious mistake.

According to the Chinese world picture, as formalized under the Han emperors, the sky was round and the Earth was square. The *Huainanzi*, a treatise on government prepared by Emperor Wu's uncle, explains, 'The Way of the Heaven is called the Round, The Way of the Earth is called the Square.'[6] Correlations could be found across nature to confirm the schema. For example, the *Huainanzi* went on to note the head's roundness resembles heaven and the feet's squareness resembles ground.[7] Likewise, the turtle could represent the universe

Mirror with square Earth, Han
dynasty, 1st–2nd century AD,
bronze with black patina.

as a whole. The circular shell on its back corresponded to the
sky, while the quadrilateral shell beneath its body symbolized
the Earth.[8] The poet Song Yu, writing in the fourth century BC,
used another analogy: 'The square Earth is my chariot and the
round Heaven my canopy.'[9]

The reverse of a bronze mirror, cast during the Han dynasty
and now in New York's Metropolitan Museum of Art, illustrates
this world picture. It is a decorated disc with a square inscribed
within a circle. The rim of the mirror portrays watery chaos
encompassing the round heaven inhabited by mythical beasts
and gods. The square represents the Earth, aligned with the
cardinal directions of north, south, east and west, with China
itself occupying the centre.

The *Classic of the Mountains and the Seas*, which reached
its final form during the Han dynasty, extended the fourfold
symmetry outwards.[10] On each side of the central lands, there
was a range of mountains, with a sea beyond. On the other side
of the waters, four great wildernesses extended to the end of

Temple of Earth, Beijing.

the world. Likewise, the *Huainanzi* says there were nine provinces in the central lands of China, arranged in a square and surrounded by four seas, beyond which there were eight other continents in the world, making nine in total. It states that each of the Earth's edges stretch for about 130,000 kilometres (80,000 mi.).[11] From this, it is clear that the Chinese envisaged a world much bigger than the modern Globe, which has a circumference of just 40,075 kilometres (24,900 mi.).

To harmonize human affairs with nature, early rulers arranged plots of agricultural land into squares as a microcosm of the whole Earth. Each plot was divided into nine smaller squares, with the one in the centre farmed by the community for the benefit of the lord. This was called the well-field system because it resembled the Chinese character for a well, and it remained an ideal long after it could no longer be implemented in practice.[12] A nine-box grid also formed the basis of town planning.¹³ Admittedly, it wasn't usually possible to arrange cities in perfect squares or farmland into equal parts. Nonetheless, important Chinese building projects, such as the Forbidden City completed in 1420, remained symmetrical, four-sided and carefully delineated according to cosmological principles. Keeping to the same theme, to the north of the capital, the Ming dynasty built a Temple of Earth, featuring a square altar. A much larger Temple of Heaven, based around the circle, is located to the south.¹⁴

The most influential book on the universe in pre-modern China was *The Gnomon of the Zhou*, compiled around AD 20, but purporting to date from centuries earlier. A gnomon is a stick used to cast a shadow, such as the one used by Eratosthenes, and was essential for timekeeping and determining the

Temple of Heaven, Beijing.

calendar. *The Gnomon of the Zhou* explains how to use the length of the shadow cast by this instrument, combined with Pythagoras' theorem (Chinese mathematicians had worked this out for themselves), to calculate the height of the Sun and how far it is from the North Pole.[15]

The Gnomon of the Zhou affirms the system of round above and square below. Heaven was shaped like a rotating umbrella or canopy 40,000 kilometres (27,000 mi.) above the Earth, with its centre over the North Pole.[16] Based on a few other assumptions and observations, it is relatively easy to calculate that the Yellow River basin in northern China should be about 53,000 kilometres (33,000 mi.) from the North Pole. The Sun orbits the pole each day, closer to it in the summer than the winter. On midwinter's day when the Sun reaches its maximum distance from the pole, it is almost 130,000 kilometres (80,000 mi.) away, making it appear lower in the southern sky from China than it does in the summer.

Because *The Gnomon of the Zhou* consisted of several disparate texts sewn together, it doesn't present an entirely consistent world picture. For instance, it implies that the North Pole, the pivot around which the heavens rotate, was also the midpoint of the Earth. This contradicts the settled Chinese view that their capital city was at the centre. *The Gnomon* is also unsure about the topography of the Earth. At one point, it states that 'heaven resembles a covering rain hat, while the Earth is patterned on an inverted pan.'[17] The bulge at the centre of the Earth means the North Pole is 32,000 kilometres (20,000 mi.) above the inhabited lands further south. This may be a remnant of an older tradition about a colossal mountain at the centre of the world, which, as we have seen, formed part of several other traditional world pictures.

One of the features of *The Gnomon*'s system is that all the heavenly bodies are on the canopy, floating 43,000 kilometres

(27,000 mi.) above us. As a result, the Sun should be visible at all hours. In some civilizations, the answer to this problem was to say that the Sun hid behind the mountain at the centre of the Earth each evening. *The Gnomon* had a different, if somewhat arbitrary, solution: the light of the Sun only extended for 90,000 kilometres (56,000 mi.), such that it becomes invisible when it is further away.[18] In the summer, when the Sun orbited the North Pole closer to China, it was visible for more of the day than when it was away to the south. The system explained how it was that days were longest in June and shortest in December. It also implied that the closer you travel to the North Pole, the longer summer days become. Once you get within about 16,000 kilometres (10,000 mi.) of it, you could expect 24 hours of daylight in the middle of the summer.

The heavenly canopy over a square earth remained the dominant model in Chinese thought until modern times, but it was not unchallenged. Zhang Heng (AD 78–139), who held the office of grand clerk under two of the later Han emperors, developed a spherical model of the universe, which was much discussed at the time. The grand clerk was responsible for seventy staff at the astronomical bureau, who submitted the annual calendar to the emperor, as well as noting important astrological portents. Zhang Heng was celebrated as an astronomer and mathematician, but also excelled as a poet and even invented a seismograph to detect earthquakes. As the epitaph on his tomb said, 'With the arts of number, he exhausted heaven and earth; as a creator, he was equal to the creative [power of nature].'[19]

He described the formation of the universe as beginning with a homogeneous mass, before continuing:

At this stage, the original substance differentiated, hard and soft first divided, pure and turbid took up different positions. Heaven formed on the

outside, and the Earth became fixed within.
Heaven took its body from the Yang, so it was
round and in motion; Earth took its body from
the Yin, so it was flat and quiescent.[20]

On this basis, Zhang Heng postulated a spherical universe, the bottom half of which was full of water. The Earth floated in the Ocean like an iceberg, the greater part of which was beneath the surface. The whole had an apparently arbitrary diameter of about 130,000 kilometres (80,000 mi.). Nonetheless, the canopy system, which the Chinese respected as venerable and straightforward, retained its dominance. Zhang Heng was not forgotten, but his lustre faded. One fifth-century commentator dismissed him, since he 'failed to understand the underlying principles, but used elaborate expressions and over-elegant explanations'.[21]

During the twentieth century, the esteemed historian of Chinese science Joseph Needham (1900–1995) brought Zhang Heng's cosmology to a wider anglophone audience. Unfortunately, Needham was defensive about Chinese ideas on the shape of the Earth. In his *Science and Civilization in China*, he quoted Zhang Heng as saying, 'The heavens are like a hen's egg and round as a crossbow bullet; the Earth is like the yolk of the egg, and lies alone at the centre. Heaven is large and Earth small.' Needham claimed that this 'clearly shows how the conception of a spherical earth, with antipodes, would rise out of [the spherical universe]'.[22] Certainly, comparing the Earth to a yolk could be read as implying it is spherical, but it is clear from other parts of Zhang Heng's writing that he had not meant to suggest this.[23] Indeed, despite his novel ideas, he didn't deviate from the established Chinese picture of a flat earth.[24]

New Religions

The Han dynasty lost the mandate of Heaven, and its empire fragmented in the early third century. However, China was still a nation of people with a shared language, shared traditions and a history that bound them together. Even today, the ethnic group from the heartlands of the old empire are called Han Chinese. There was a powerful centripetal force towards unity, but it took three centuries after the fall of the Han to overcome the fissiparous inclinations of local rulers and pull China back together. In 618, a new Tang dynasty reunited all under Heaven. The phrase 'all under heaven' meant the central lands themselves, although it hinted that the rest of the world owed tribute to the emperor as well.

Chinese culture had evolved since the Han era. A lasting change was the introduction of Buddhism from India. Although Buddhism would all but die out in India itself, missionaries spread the faith across central and Southeast Asia, as well as to the islands of Sri Lanka and Japan. They made many converts in China, where it joined Taoism and Confucianism as one of the major ethical systems. By the start of the Tang dynasty, Buddhists were well ensconced in Chinese society, albeit treated with suspicion by some of the learned elite. Missionaries continued to arrive from India, and Chinese monks set off in the other direction to collect copies of Buddhist and Sanskrit texts from the libraries of the Ganges Basin.

Among the Indian books brought to China were manuals on astronomy. So, it made sense that when the imperial calendar needed adjustment (again), officials looked to Buddhist monasteries for the personnel to sort the problem out. In AD 721, a monk called Yixing (683–727) was summoned by the throne to report on how the calendar could be reformed. The emperor had been getting agitated that it was no longer possible to accurately predict solar eclipses. While he had some familiarity with Indian astronomy, Yixing relied on conventional Chinese sources for his calendar, which was used for the period from 729 to 761. He was more innovative in his construction of an astronomical instrument called an armillary sphere. This was a framework of rings that represented the paths of the Sun, Moon and other aspects of the heavens. Yixing added the ecliptic, which had always featured on Greek models, although earlier Chinese versions do not seem to have included it. This enabled him to show how the paths of the Sun and Moon intercept when there is an eclipse.[25]

The emperor also commanded Yixing to carry out a survey of China. In particular, he needed to locate the Earth's centre, since this was where certain rituals should best be carried out. Rather than being at the North Pole, as we might expect from *The Gnomon of the Zhou*, it was traditionally situated at the city of Dengfeng, one of the ancient capitals of China. Yixing relocated it about 195 kilometres (120 mi.) east to Kaifeng, another antique capital, on the south bank of the Yellow River. He could have justified this move by studying the shadow cast by a gnomon. One of the geographical problems with a flat earth was finding the direction of true north. On the Globe, that's easy, at least in theory. Wherever you are, the shadow of the Sun at midday is at its shortest and always points due north or due south, unless the Sun is directly overhead and there is no shadow. But this doesn't work if the Earth is flat. In that case, the shadow only points exactly north at midday if you are standing on a line running directly south from the North Pole. Chinese astronomers were acutely aware of the difficulties of accurately measuring the shadow of the gnomon, and how this made it hard to determine the exact whereabouts of the Earth's centre.[26]

In AD 730, after Yixing had died and his calendar had come into use, he was accused of plagiarizing an Indian manual of astronomy that had been translated into Chinese a few years previously. This book, called the *Navagraha-karana*, described a cosmological system descended from the spherical astronomy of Aryabhata, with the adaptations needed for the latitude of the Tang capital city. Thus, it is one of the earliest works in Chinese that assumed the Earth is a sphere. Although he was aware of the theory of the Globe, Yixing didn't use it for his calendar and the theory didn't catch on. Later Tang astronomers lacked a background in the concepts used in the *Navagraha-karana* and so dismissed it as inaccurate.[27] Other aspects of Indian astronomy were more influential, in particular the dark star Rahu. This was the invisible entity hypothesized to explain eclipses in traditional Indian astronomy. Now, the idea came to China as well, rivalling the true explanation that eclipses occurred when the Moon passed in front of the Sun.[28]

At its peak, the Tang Empire extended its territory far into Central Asia, where it encompassed congregations from the Christian Church of the East as well as Zoroastrians and Manichaeans. We've already examined the remarkable Manichaean *Diagram of the Universe* and the Christians' tabernacle world picture. The Church of the East transmitted Western manuals of astrology to China at around this time as well, but these didn't lead to the Aristotelian world view leaking through to the local intelligentsia.[29] On the contrary, an inscribed stele, carved in the eighth century, gives a hint that these Christians adopted the traditional Chinese world picture. The stele celebrates their success under the Tang dynasty, as well as making plain their obsequious loyalty to the emperor. Written in Chinese, the stele includes a summary of the Christian creed. It begins:

Dividing the cross, [God] determined the four cardinal points. Setting in motion the primordial spirit, he produced the two principles of nature. The dark void was changed, and Heaven and Earth appeared. The sun and the moon revolved, and day and night began.[30]

While this is a paraphrase of the creation account in Genesis, it incorporates Chinese elements. For example, the two principles of nature refer to yin and yang, while the association of the cross with the cardinal points correlates to the Chinese picture of the Earth with sides aligned north, south, east and west. Christians' accommodation to Chinese ways of thinking was not sufficient to guarantee their safety, and they were proscribed in AD 845, together with Zoroastrians and Buddhists. The ban on the Buddhists was rescinded just two years later, but Christianity had to keep beneath the notice of official surveillance and eventually died out.[31] The stele survived only because it was buried to keep it safe.

The Struggle with the Tribes

Wracked by rebellion, Tang China became increasingly autarkic through the ninth century as the mandate of Heaven began to dissolve. A few decades of strife followed the dynasty's fall before the Song took power in 960. Kaifeng, now recognized as the centre of the Earth, was chosen for the capital.

The Song era was when China most obviously put to shame the state of European civilization. In the eleventh century, blast furnaces burned coke rather than wood to smelt over 100,000 tons of iron a year, a rate of production that Europe reached only in 1700. Meanwhile, farmers developed new hybrid strains of rice to feed the burgeoning population.[32] Devices similar to the cotton-spinning 'Jenny', invented in England in the 1760s, existed in China in the fourteenth century but did not lead to an industrial revolution.[33] Perhaps the surge in the labour force was why mechanization did not catch on. The Song used paper money to ensure economic growth was not restricted by the availability of bullion, although the government was unable to prevent the rampant inflation that resulted.[34]

Song China was under constant pressure from tribes to its north and west. These groups gradually pushed south, taking over large swathes of the territory once ruled by the Tang. The Song capital at Kaifeng was captured in 1127. The dynasty regrouped and continued to rule southern China for another century and a half as the north fell to invading Mongols in 1234. Kublai Khan (1215–1294), a grandson of Genghis Khan (1162–1227), founded the Yuan dynasty when he proclaimed himself emperor in 1271. He continued to push south and completed the conquest of the Song in 1279. By this time, the Mongols ruled a vast swathe of land from eastern Europe to the Pacific, including Central Asia, the Middle East and Persia.

While Kublai Khan occupied himself subjugating China, his brother Hulagu (1217–1265) was crushing resistance to Mongol rule at the other end of Asia. In 1258, Hulagu's armies sacked Baghdad, razing the city and massacring its inhabitants in a week-long orgy of pillage. Two years previously, he had captured Alamut, the fabled castle of the Assassins, without a fight. Despite his penchant for murderous rampages, Hulagu was interested in astrology, so housed the astronomers he captured from the Assassins in a new observatory at Maragha in northern Persia. It's likely that he sent one of his menagerie of savants, a Persian Muslim called Jamal al-Din (*fl.* 1267–88), to his brother Kublai with a package of astronomical instruments, including dials, an armillary sphere and an astrolabe.[35] Jamal also carried a terrestrial globe – the first time one of these is recorded in China.[36]

Kublai Khan acknowledged that his empire was home to a substantial Muslim population. To cater for their need for a calendar, he founded an Islamic astronomical bureau in his capital and appointed Jamal al-Din as its director.[37] The Muslim and Chinese astronomical bureaus existed in parallel for several centuries, but there does not seem to have been much cross-pollination between them. So, while we might expect that communication between Persia and China would lead to the acknowledgement of the spherical earth by Chinese astronomers, this didn't happen. They took no notice of Jamal al-Din's terrestrial globe nor the methods of Arabic astronomy, which was an improved version of Ptolemy's Greek original. In 1288, members of the Chinese astronomical bureau even attempted to get Jamal al-Din dismissed.[38]

Furthermore, when Kublai Khan wanted a new calendar to mark the beginning of his dynasty, he knew it had to be acceptable to his Chinese subjects. Ignoring the Muslim bureau, he instructed a leading native astronomer, Guo Shoujing (1231–1316), to formulate it. Guo Shoujing was a brilliant instrumentalist and mathematician who, despite working within the traditional paradigm, produced a calendar that was more accurate than the Islamic one developed next door by Jamal al-Din.[39] Indeed, there is no proof that Guo Shoujing's work was at all influenced by Western developments. Instead, his calendar was a triumphant reassertion of the superiority of the Chinese way of doing things.

Outwards and Inwards

In the mid-fourteenth century, the Chinese rebelled against their Mongol overlords and, after a period of civil war, the new Ming dynasty was installed in 1368. The first Ming emperor oversaw a return to Confucian values after the Mongol interim. The reign of his son, the Yongle Emperor (1360–1424), was characterized by a self-confidence that he wanted to project beyond China's borders. This was the context for the famous voyages of Zheng He (1371–1433), a Muslim captured as a child, castrated and taken into royal service. In 1405, the emperor gave Zheng He command of an enormous fleet of junks to spread news about his benevolence to the known world. With 317 ships and almost 28,000 men at his disposal, the admiral had little trouble overawing the countries where he stopped off. He distributed largesse to demonstrate the emperor's munificence and gathered gifts to take home, which he termed as tribute. These included a giraffe from the east coast of Africa, which the Chinese took to be a mythical beast that embodied good fortune.[40] When the little ships of the Portuguese expedition of Vasco da Gama pulled into the same shore eighty years later, the tradition of the visitation from China was still remembered.[41]

In all, Zheng He led five voyages between 1405 and 1433 before a change in policy and emperor led to their discontinuation, most likely because they were ruinously expensive.[42] The money was needed elsewhere. Northern tribesmen were again threatening all under heaven and captured the emperor himself in 1449. This led to retrenchment by the Ming, who abandoned any foreign adventures and fortified the Great Wall.

Zheng He was not engaged in exploration. He knew exactly where he was going and how to get there. Merchants had plied the Indian Ocean for over a thousand years, taking advantage of the seasonal variations of its monsoonal winds. We've already come across the Roman traders who sailed between the Red Sea and India, perhaps carrying knowledge of the Globe with them. When ships crossed the open ocean, they expected to be blown by a following wind towards a coast they already knew about. They navigated along bearings from one port to another that they maintained by observing the stars. Zheng He never had any intention of visiting unknown lands, not least because he assumed all civilized people had already been discovered. His fleets spread Chinese influence and helped to control trade, but there was no motivation to expand the borders of the empire into overseas territories.

If anyone had been looking for it, Zheng He's far-flung travels could have provided plenty of evidence for the curvature of the Earth. He travelled far enough south to lose sight of northern stars and to reveal those only visible towards the Antarctic. More obviously, every time his junks sailed away from the shore, the crew would have been able to watch coastal landmarks sink beneath the horizon. Chinese cosmology could account for both of these observations. As we've seen, the model of the heavens in *The Gnomon of the Zhou* supposed that the stars fade from sight once you get too far away from them, which would explain the disappearance of Northern Hemisphere stars once the observer crossed the equator. As for the horizon, the *Gnomon* did concede that the Earth, while essentially a plane, might bulge upwards towards the North Pole, giving its surface some curvature. This happened to mean that China itself, being nearer the pole, was elevated relative to the rest of the civilized world and also explained why major rivers flowed downwards into the four seas.[43]

There was another aspect of Chinese thought that dissuaded Zheng He from reconnoitring new lands for the sake of it. China itself was at the centre of the world and the pinnacle of civilization. Under the Ming, it first took the familiar moniker of 'the Middle Kingdom'.[44] The further from China that Zheng travelled, the less civilized he expected the locals to become. Beyond the bounds of the known world there was only barbarism and poverty, so there really was no point in venturing that far. And since the emperor already ruled all under heaven, titularly if not substantively, conquest was extraneous as well. It was not as if these faraway people were a threat, unlike the marauding horsemen of the northern steppe. And yet, less than a century after Zheng He returned from his final voyage, some of these distant barbarians from beyond the edge of the map turned up in southern China and started to make a nuisance of themselves.

21

China and the West: 'Like the yolk in a hen's egg'

Portuguese traders began to enter China in the early sixteenth century. They were so belligerent that the Ming authorities made repeated efforts to throw them out but, in 1577, the government leased the foreigners a small piece of land at the mouth of the Pearl River. The intention was to contain the Portuguese, while benefiting from the commerce they brought. Shortly afterwards, a few Jesuit missionaries slipped into China with the ambition of converting the Middle Kingdom to Christianity. The Society of Jesus had been founded by St Ignatius of Loyola (1491–1556) in 1540, and quickly found its groove as an educational and proselytizing order. Jesuits accommodated themselves to the cultures they worked in and did not have to wear priestly vestments. In China, they initially dressed as Buddhist monks before switching to the robes of the learned mandarin class.

The Chinese treated the Jesuits as an amusing novelty while intermittently persecuting them. They admired the Europeans' knowledge but found some of their habits, like self-imposed celibacy, deeply disconcerting. Overcoming these suspicions, the Jesuits curried favour with the mandarins and the emperor, while scorning other elements of Chinese society like the court eunuchs. Ultimately, despite a century and a half of effort, the Jesuits' mission was a failure. Although they nurtured a Christian community that eventually ran into the tens of thousands, this was insignificant compared to the country's vast population.

Removing the Middle Kingdom from the Middle

As part of their evangelizing effort, the Jesuits employed Western science and technology to make themselves indispensable to the imperial authorities. In return, they sought permission to preach freely. The strategy was spearheaded by an Italian priest called Matteo Ricci (1552–1610), who arrived in China in 1582. Cultured and gregarious, he mastered the written language and nurtured contacts within the literate elite. His ultimate aim was to reach the emperor and gain official sanction for missionary work.

Ricci knew that Catholicism and Aristotelian cosmology would appear to be dangerous novelties to the Chinese, so he hit on the plan of sugar-coating his message in the language of the Confucian Classics. This played to the conservatism of his Chinese audience. Ricci had to convince them that they had strayed from the path ordained by their ancestors and that he could use science to guide them back to the ancient texts' true meaning. Geography gave him an early opportunity to put his scheme into effect.

Shortly after he arrived, Ricci noticed Chinese visitors to the Jesuits' residence gawping at a map of the world hanging on the wall.[1] He knew why they found it so strange. In letters he wrote in late 1595, he affirmed that the Chinese thought the Earth is flat and square, and the sky a round canopy. They also thought that there was a vacuum in space, which contradicted the European view that the heavens consisted of spheres composed of an incorruptible fifth element. And while the Chinese understood the cause of solar eclipses, they imagined that the Sun hid behind a mountain at night.[2] He was dismissive of these views in private, but Ricci knew he needed to be courteous about how he sought to correct them. He decided to produce a world map for a Chinese audience.

P·MATTHEVS RICCIVS MACERATENSIS QVI PRIMVS E SOCIETAE IESV EVANGELIVM IN SINAS INVEXIT OBIIT ANNO SALVTIS 1610 ÆTATIS, 60

You Wenhui (alias Manuel Pereira), *Matteo Ricci of Macerata*, c. 1610, oil on canvas.

The Chinese had excellent maps of their country, so Ricci combined these with European charts plotted during the voyages of the Portuguese and Spanish. Diplomatically, he placed the Americas on the right and Europe on the left, so China remained near the centre. He filled out the unexplored parts of the world with elements of Chinese and Western fantasy, such as the land of the dwarfs and a realm of the one-eyed people.[3] The Southern Hemisphere was dominated by an enormous and non-existent continent, *Terra australis*, that Europeans had convinced themselves was awaiting discovery.

Ricci improved his map in several stages before it reached its most developed form, which he called the 'Complete Geographical Map of Ten Thousand Countries'. When printed, it was 4 metres (12½ ft) long and over 1.5 metres (5 ft) high.[4] It was more than just a representation of the continents. Ricci included diagrams of the whole universe in each corner, as well as lengthy explanatory captions. He broached the theory of the Globe in the map's general introduction:

> The Earth and sea are both spherical. Together they form a single orb situated at the centre of the celestial spheres, like the yolk in a hen's egg that is surrounded by the white. Those who said the Earth is square were referring to the Earth's fixed and immobile nature and not its physical form.[5]

This passage is a paradigmatic example of the way Ricci reinterpreted ancient Chinese texts to reflect European cosmology. When Zhang Heng had used the analogy of the egg and yolk in the second century, he was not proposing that the Earth was a sphere, merely that the heavens surround the Earth on all sides. That didn't stop Ricci from appropriating the metaphor and applying it to Aristotelian cosmology. He also reinterpreted statements that the Earth is square to make them figurative rather than literal.

In religious matters too, Ricci had the audacity to imply he understood Confucian literature better than the Chinese did. The most notorious example of this was his reinterpretation of the word 'Heaven' to make it more conducive to Christianity. In Chinese tradition, Heaven was an impersonal force, not dissimilar to the moral order that the ancient Greeks imagined governed their lives. Ricci equated the Christian God with a 'Lord of Heaven' and implied that this was consistent with the message of the Confucian Classics. The Jesuits called their Chinese catechism 'The True Meaning of the Lord of Heaven', which sought to distil, or rather create, a pure form of Confucianism that was compatible with Christianity.[6] For good measure, Ricci also denigrated the tenets of Buddhism and Taoism.

Scholars working under the Song dynasty had modified Confucian doctrine to make it more apposite for running the secular Chinese state. The resulting philosophy, which had become the basis of the civil service exams under the Ming, is nowadays called neo-Confucianism.[7] Ricci disparaged neo-Confucianism as a late accretion to the tradition. Unfortunately, many of the mandarins, though they admired him, could see exactly what he was up to. Things got especially fraught when Ricci started to talk about the virgin birth and the incarnation of God as a man in Jesus. 'Our Confucianism has never held that Heaven had a mother or a bodily form,' one friendly Chinese critic complained later.[8]

The Jesuits finally succeeded in penetrating the Forbidden City in 1601 and maintained a presence there for much of the next 150 years. However, they didn't make much headway in their efforts to convert the Chinese elite to Christianity, let alone the emperor. There were a few exceptions, two of whom were Xu Guangqi (1562–1633) and Li Zhizao (1565–1630). Both accepted Catholicism under Ricci's influence. Xu helped the Jesuits translate European works into Chinese, all the while mastering Western methods himself. Li prepared the Chinese translation of Aristotle's *On the Heavens*, published in 1628.[9]

Once they were in the Forbidden City, the Jesuits hoped their expertise in astronomy would make them so essential that their missionary activities would be officially tolerated. They seemed to be making headway but began to lose influence following the death of Matteo Ricci in 1610. A change in government in 1616 saw them expelled from the capital and exiled to Macau. Across the rest of China, they were forced into hiding for several years.[10]

Xu Guangqi and Li Zhizao worked hard to rehabilitate their Jesuit friends, eventually succeeding in convincing a new emperor that the calendar was in urgent need of reform.[11] The Christians also received aid from Heaven itself. In 1629, there was a solar eclipse over the capital. The incumbent Chinese astronomers and the Jesuits both made predictions about the occasion. The former's calculation of the time it would start erred by an hour, and they said it would last two hours instead of the actual duration of just two minutes. The Jesuits were correct on both counts.[12] As a result, the emperor ordered them to participate in the development of a new calendar to mark his reign. It took until 1642 to complete, by which time the politics of China looked very different.[13]

Enter the Qing

The mandate of Heaven was already leaking away from the Ming dynasty. From 1618, Manchu tribes from the northeast inflicted a series of defeats on China that would culminate in the capture of Peking in 1644. The Manchu instigated the new Qing (or Ch'ing) dynasty, but resistance from those loyal to the old regime lasted for decades. The Jesuits tried to hedge their bets, but once it was clear the Qing had won, they had little hesitation transferring their attention to the new dynasty. In 1645, a German Jesuit, Johann Adam Schall von Bell (1591–1666), curried favour with the Qing emperor by presenting him with the new calendar, deftly repurposed to veil its Ming origins. In return, Schall von Bell was appointed as director of the astronomical bureau, a position that would be held almost exclusively by Jesuits until 1775.[14]

The Christians were now in charge of the calendar, timekeeping and astrological interpretations. The latter was especially problematic for them, since it involved determining auspicious dates for traditional Chinese rites that Catholics were supposed to abhor. The Jesuits claimed these rites were secular in nature, rather than religious, so that assisting the imperial government to ensure they took place at the right time didn't involve condoning superstition. Still, mandarins opposed to the Jesuits could be forgiven for thinking they did not treat the traditional ceremonies with the seriousness they deserved.

As long as he had the favour of the emperor, Schall von Bell was safe. But he was careless about making enemies. In 1657, he fired the Muslim staff at the astronomical bureau, severing the connection with Islam that had existed since the reign of Kublai Khan in the thirteenth century.[15] Wu Mingxuan (*fl.* 1657–69), one of the sacked Muslim astronomers, tried to retaliate against the Jesuit by accusing him of making inaccurate predictions. When the charge was dismissed, Wu was thrown into prison. On his release, he teamed up with a mandarin called Yang Guangxian (1597–1669), who had got himself into trouble for throwing false allegations at senior ministers. While Wu was a professional astronomer, Yang's talents were literary. He wrote a series of memorandums attacking the Jesuits for their cosmological novelties, which he showed were incompatible with the traditional Chinese calendar.[16]

Among other matters, Yang challenged the theory of the Globe. One of his arguments was that it was absurd that the seas don't pour away from a spherical earth, or at least gather at its base. 'If indeed there are countries existing on the curved edge and the bottom of the globe,' he explained, 'then these places are surely immersed in water. Westerners [living on the other side of the world] then must surely belong to the likes of turtles and fishes.' Instead, he reiterated the model of Zhang Heng – that the Earth floats on water inside a spherical universe. 'Since the Earth resides on the water,' he said, 'it is evident that all ten thousand countries are located above the horizon ... for the horizon is nothing but the level surface of the water of the Four Seas.'[17]

At first, Yang's memorandums were knocked back before any senior minister had seen them. Nonetheless, the Jesuits foolishly rose to Yang's bait and denounced him as disloyal to the new Qing dynasty. He retaliated by accusing the Europeans of choosing an inauspicious time for the burial of an infant prince who had died in 1657. This was a serious charge. The dead had to be interred in the right place and at the right time. Failure to observe these rules would lead to misfortune for members of the family who were still living. In this case, it appeared that retribution had been swift: the first Qing emperor and his wife had both died of smallpox in 1661.[18] Given their heir, the Kangxi Emperor (1654–1722), was aged just eight, a regency of four mandarins ruled the country until he came of age.

In 1664, Yang forwarded his complaint about the mistimed burial to the Ministry of Rites. This time, his accusations stuck. The Jesuits and their Chinese colleagues in the astronomical bureau were carted off to prison to await trial. Their mistreatment while in jail was too much for the ailing Schall von Bell, who suffered a stroke. Then, in April 1665, he was sentenced with his colleagues to death by dismemberment. It looked like only a miracle could save them.

The day after the trial, an earthquake struck northern China and damaged the Forbidden City. It was axiomatic to the Chinese that natural disasters were warnings from Heaven about misgovernment or unjust punishments. To be on the safe side, the sentences against the Jesuits were remitted to house arrest, where Schall von Bell died of ill health shortly afterwards. Five unfortunate Chinese astronomers who had collaborated with the Europeans were beheaded.[19] Yang suddenly found himself triumphant. The government ordered him to take over the running of the astronomical bureau, which was the last thing he wanted. He protested that he didn't know anything about mathematics and wasn't capable of compiling the calendar. Compelled to take the post, he installed Wu, his Muslim co-conspirator, who was at least a trained astronomer, as deputy director.[20] Unfortunately, he was unable to match the accuracy the Jesuits had achieved.

In 1668, the young Kangxi Emperor started to intervene in affairs of state and judged the calendar fiasco a good way to test his mettle against the regents. He summoned Ferdinand Verbiest (1623–1688), now the senior Jesuit astronomer following the death of Schall von Bell, and asked him to review the almanac Yang and Wu had produced for the year. Predictably, Verbiest found numerous mistakes. The emperor responded by demanding a contest between the Jesuit and Yang to predict the height of the Sun and other astronomical observations. These tests were carried out with instruments specified by Verbiest and in accordance with Western concepts of the heavens. It's likely the emperor intended that the Jesuits should win as a way to start wresting power from the regents and taking command of the government himself. In any case, Verbiest manipulated the results and was easily able to convince a committee appointed by the emperor that his predictions were more accurate than those of the traditionalists. Yang was sacked as director of the astronomical bureau and sent home in disgrace. Verbiest replaced him, putting the Jesuits back in the saddle. Wu kept his job for a little longer before he was flogged for incompetence.[21] In time, Chinese conservatives would revere Yang as a martyr, applauding his opposition to the foreigners who had sought to subvert tradition and introduce alien rites.[22]

The Jesuit mission to China was eventually stymied by the Catholic authorities in Europe. In the early eighteenth century, the papacy decided that accommodation to Chinese mores had gone too far and sought to ban Christians from taking part in rites such as the veneration of ancestors. Efforts by the papal legate to China to enforce this edict offended the emperor and led to the expulsion of the missionaries in 1724, when Christianity was banned outright.[23]

Ambivalence about the Globe

The Jesuit scientific project was supposed to have been one of openness. They translated European scientific works into Chinese and expected that these would quickly supersede traditional texts like the *Gnomon of the Zhou*. By advertising the superiority of Western civilization, they hoped to clear the way for the conversion of China to Catholicism. The Kangxi Emperor had his own agenda. It suited him to have the astronomical bureau run by Jesuits, whose presence in the Forbidden City was entirely at his own pleasure. But he had no intention of publicizing the methods they used. Astronomy, after all, was an imperial prerogative. He sent Manchu loyalists, members of his own tribe from north of China, for training at the bureau and personally took lessons in Western science from Verbiest. Meanwhile, he ordered that astronomy should not feature in the civil service examinations that all potential mandarins had to pass, thereby ensuring the subject was excluded from the syllabus in Chinese schools.[24]

As a result, the theory of the Globe didn't become common knowledge or supplant the traditional world picture. Nor was it included in official compilations of learning. In the 1770s, a group of 360 lexicographers undertook a major project at the imperial library to catalogue every volume printed in China. They evaluated and summarized the books, excerpting them into a universal encyclopaedia. By this time, there were Chinese treatises that mentioned the Globe, such as a 1648 textbook that includes a diagram of a round earth with a pagoda on one side and a cathedral on the other.[25] Yet, the concept was entirely absent from the imperial library catalogue, even though it incorporated manuals on technical European disciplines.[26]

The shape of the Earth continued to be a matter of debate among the intelligentsia through the seventeenth and eighteenth centuries. Some Chinese scholars thought the world must have an irregular form that was neither flat nor round. One writer suggested the mottled pattern on the Moon was a reflection of the Earth's indeterminate appearance.[27] Ruan Yuan (1764–1849), among the most competent Chinese mathematicians of

Globe from a Chinese astronomy textbook: Xiong Mingyu, Gezhi Cao 格致草 (Draft for Investigating Things and Extending Knowledge, 1648).

the early nineteenth century, was sceptical about both the Globe and Copernicus's theory that the Earth orbited the Sun.[28] The Jesuits had unambiguously introduced heliocentrism to China after the general prohibition of the subject had finally been rescinded by the Catholic Church in 1758.[29]

Some Chinese scholars did accept elements of European astronomy and mathematics, including the spherical earth. One example was the polymath Mei Wending (1633–1721), who came from a distinguished family of Ming loyalists. This pedigree prevented him from holding a job at the Qing astronomical bureau but he nonetheless attracted the attention of the Kangxi Emperor, who was on the lookout for native Chinese prowess in the sciences. Mei urged the adoption of Western methods but insisted that they had their origin in China.[30] To justify this, he sought to assimilate the histories of Western and Eastern science. As we've seen, Matteo Ricci had humoured the idea that ancient Chinese texts supported the theory of the Globe when he resuscitated the spherical heaven of Zhang Heng, albeit ignoring the bits where Zhang had said the Earth is, in fact, flat. Likewise, Mei exploited a creative reading of the Confucian Classics to claim, 'The westerners' instruments for observing the celestial phenomena, the theory of the five climate zones, the idea that the Earth is round . . . none of these goes beyond the ground covered by the *Gnomon of the Zhou*.'[31]

To explain how ancient Chinese science had migrated to Europe, Mei suggested that when the legendary Emperor Yao had sent out his four astronomers to determine the times of the equinoxes and solstices in 3200 BC, the one sent west kept going, carrying his knowledge of the heavens with him. Mei explained that later, during the Warring States period of the fourth and third centuries BC, which preceded the unification of China under the Qin dynasty in 221 BC, more astronomers fled to avoid the ruinous civil strife. Some found their way to

Arabia and Greece, seeding the local scientific traditions. The Kangxi Emperor found it convenient to deprecate Western science by supporting such speculation about its Chinese origin.[32] It remained a commonplace among mandarins into the twentieth century, even after the theory of the Globe had finally won widespread acceptance in the 1800s.[33]

China suffered terribly from the depredations of the European powers but was never formally colonized. Following the era of exploration, much of the rest of the world fell under Western domination until the mid-twentieth century. Knowledge of the Globe spread with colonial rule as imperialists sought to inculcate their 'Enlightenment values', but even in Europe there were some who still hankered after a flat earth.

22

The Globe Goes Global: 'At the round Earth's imagined corners'

Christopher Columbus and Ferdinand Magellan had attempted their voyages because they were ignorant about the size of the Earth and the position of the Americas. It took centuries for the correct picture to emerge. To begin with, navigators needed an accurate way to determine their latitude and longitude. Determining latitude was the easier of the two. In the Northern Hemisphere, one way is to measure the elevation of the pole star, Polaris.[1] Alternatively, you can observe the height of the Sun at midday, which works in both hemispheres as long as you also know the date.

To calculate the size of the Earth requires additional information. As Eratosthenes had deduced, you need the exact co-ordinates of two points on the globe and the distance between them. Once you know the length of one degree of latitude, you can multiply it by 360 to give the circumference of the Earth around the poles. In early modern Europe, the best way of measuring long distances was triangulation. This technique involves measuring the length of a baseline, then finding the bearing of a landmark from each end of the baseline to form a triangle. Trigonometry can be used to work out the lengths of the other two sides of the triangle. These lengths are the new baselines to determine the distances to other landmarks. All the triangles eventually build up into a map showing the landmarks properly spaced. This requires considerable resources, as well as meticulous measurements, so it was the kind of project that only governments could realistically afford.

Determining longitude so you can accurately calculate the width of the Atlantic or Pacific is a good deal trickier. The theory, at least, is simple. If you know the time of day in Greenwich in southeast London, where longitude is zero by definition, and the time of day where you are standing, you can easily work out your longitude. It will be one degree for every four minutes' difference in the time between your location and at Greenwich. That's because the Earth rotates through 360 degrees every 24 hours, that is, every 1,440 minutes. So, every four minutes of time difference corresponds to one degree of rotation.

Back in the sixteenth and seventeenth centuries, it was hard to know the time in London while you were in America. It wasn't possible to ship a pendulum clock across the Atlantic and expect it to accurately keep time along the way. Luckily, if a lunar eclipse was visible from both continents, it provided an absolute synchronization.[2] The Spanish government was the first to launch a systematic campaign to measure longitude using eclipses of the Moon, starting with one on 27 September 1577. At least on land, this method allowed for an accurate map of Spain's empire. However, it was no use for travellers, as lunar eclipses are uncommon events. The British clockmaker John Harrison (1693–1776) solved the problem by inventing a chronometer that kept time at sea. This allowed sailors to carry Greenwich time with them and compare it directly to their local time. Eventually, in the nineteenth century, the invention of the telegraph meant that local times could be transmitted across vast distances instantaneously.

Is the Earth Really a Sphere?

Until the late seventeenth century, European scholars assumed that the Earth was a perfect sphere, notwithstanding the insignificant wrinkles from mountains and valleys. Then, in 1671, an expedition set out from France to the town of Cayenne, on the north coast of South America, to carry out observations of stars not visible in Europe. The expedition was equipped with state-of-the-art pendulum clocks that had been calibrated in Paris to ensure they were accurate. When the clocks were wound up and restarted in Cayenne, they lost 2 minutes 28 seconds a day.[3] Since pendulums work by gravity, this meant that the gravitational pull in Cayenne was slightly weaker than in France. Word of the result reached the English mathematician Isaac Newton, who was working on his theory of gravity in Cambridge. He included the discovery in his *Mathematical Principles of Natural Philosophy*, published in 1687.[4] In the book, he predicted that the Earth wasn't uniformly round but slightly squashed at the poles.

French mathematicians at the time found Newton's ideas unconvincing. They preferred the theory that the Earth ought to be narrower at the equator and tapered slightly at the poles: the opposite of Newton's prediction. To resolve the dispute, a team set out from France in 1735 on an astoundingly ambitious expedition to measure a degree of latitude at the equator. The location chosen was near the city of Quito in Peru. Another group travelled to the Arctic Circle to perform the same task. When the verdicts were in, it was clear that Newton was right: a degree of latitude is slightly shorter at the equator than at the poles.

We can now determine the shape of our planet with the astounding levels of accuracy required by modern GPS systems. The figure of the Earth, as this shape is called, is an oblate spheroid about one part in three hundred wider at the equator than it is tall. In other words, its equatorial diameter is about 42 kilometres (26 mi.) more than it is from pole to pole. This is slightly less squashed than Newton predicted. In comparison, the deepest ocean (11 kilometres/7 mi.) and highest mountain (9 kilometres/5½ mi.) are smaller blemishes. The bulge on the equator means that the spot on the Earth's surface furthest from its centre (rather than the highest above sea level) is at the top of a mountain in Ecuador, South America.

Evangelizing the Globe

As Europeans plied the oceans, they gradually filled out their maps with every coast and island. However, one thing eluded them. After spending 2,000 years arguing about whether the antipodes existed, by the eighteenth century they had convinced themselves there was a great southern continent, dubbed *Terra australis*. Plenty of explorers thought they had discovered it, or at least seen it from afar. Besides, there seemed to be an awful lot of southern ocean without very much in it. It wasn't until the late eighteenth century that Captain James Cook (1728–1779) finally dispelled the myth of *Terra australis* by sailing around the Antarctic Circle without encountering it.[5] Of course, there is a southern landmass: Antarctica, which was discovered in the 1820s. This completed the map of the seas, even though parts of the interiors of the continents remained uncharted for many more years.

In Europe, terrestrial globes became a common item of furniture in the salons of the wealthy. Yet, as Europeans continued to explore and colonize the world, they found belief in traditional world pictures was widespread. For instance, antiquarians, studying the remains of the Aztec and Maya cultures that had survived the looting of Spanish conquistadors, found complex astronomical systems that nonetheless contained no hint of the Globe. The Aztecs centred the Earth on their capital of Tenochtitlan, which was itself an island in a saline lake, making it a microcosm of the whole world. The Earth was surrounded by four petals, corresponding to the four cardinal points. The southern petal was a paradise, ruled by the rain god. In the north, there were entrances to the underworld, through which the Sun had to pass each night before rising in the east. The sky consisted of thirteen layers, each the home of various gods.[6]

To the south of Tenochtitlan, the Maya shared many of the Aztecs' assumptions about the universe, except that they associated the north with life-giving rains and the south with death. Branches of a cosmic tree held up the sky, with its roots in the underworld.[7] Across the Pacific, Polynesians and Aboriginal Australians had their own world pictures. In Hawaii, people thought the sky was a rotunda shaped like a gourd.[8] The Tipi people of Australia described the Earth as a disc, surrounded by water and covered by the dome of the heavens.[9]

With all this information flowing into Europe, the founders of the new subject of anthropology took it for granted that the Earth's flatness was something everyone naturally assumed until they had learnt otherwise. As Edward Burnett Tylor (1832–1917), the first professor of anthropology at the University of Oxford, wrote:

> Children living unschooled in some wild woodland would take it as a matter of course that the Earth is a circular floor, more or less uneven, arched over with a dome or firmament springing from the horizon. Thus the natural and primitive notion of the world is that it is like a round dish with a cover. Rude tribes in many countries are found thinking so, and working out the idea so as

to account for such phenomena as rain, which is
water from above dripping in through holes in the
sky-roof. This firmament is studded with stars,
and is a few miles off.[10]

Tylor thought this was as true for villagers in rural England
as it was for what he called 'rude tribes'. A few years previously,
he assured his readers, a schoolteacher lecturing on the Globe
in the West Country had provoked to fury the country folk who
had thought the Earth was flat.[11]

Europeans like Tylor treated traditional world pictures as
quaint or primitive. Even so, few were as condescending as
Thomas Babington Macaulay (1800–1859), who later became
famous for his *History of England*. He started his career as a civil
servant in British India, where he took part in a discussion over
the allocation of funds for schools. Parliament in London had
insisted that the Indian administration spend money on educating native peoples, which the colonial authorities took to mean
supporting Hindu and Muslim colleges. Macaulay forcefully
rejected this course, lauding the superiority of English science
and scholarship. In his *Minute on Education* of 1835, he dismissed
Hindu learning as composed of 'Astronomy, which would move
laughter in girls at an English boarding school; History, abound-
ing with kings thirty feet high, and reigns thirty thousand years
long; and Geography, made up of seas of treacle and seas of
butter'.[12] The Hindu and Muslim colleges that Macaulay denigrated were effectively seminaries for Brahmins and imams, so
his assessment of their curriculums was like dismissing Western
science on the basis that the Book of Genesis was studied at theological colleges. He was not remotely interested in the efforts we
encountered earlier to reconcile the Globe to the Hindu Puranas.
Nonetheless, Macaulay won the argument and the funding was
diverted to teaching European subjects to Indian people.

For Macaulay, traditional world views were a matter of academic interest. For missionaries travelling around Asia and Africa to spread Christianity, Western cosmology was an irresistible way to cast aspersions on the beliefs they were attempting to replace. Unperturbed that this strategy had had limited success for Jesuits in China, Protestant evangelists were happy to try it out as well. As the English divine and poet John Donne (1572–1631) had sung, trumpets should blow for the Lord 'at the round Earth's imagin'd corners'.[13]

Colonial Missionaries

In August 1873, two men – a Buddhist monk and a Protestant clergyman – stood before 5,000 people outside the city of Colombo in Sri Lanka. They were there to engage in a debate, set to last two days, over the truth of their respective religions. The Christian had placed a table on his side of the platform, covered in a white tablecloth and decorated with some greenery. Ranged against him were two hundred Buddhist monks in saffron robes beneath a canopy of red, white and blue.

In his opening speech, David da Silva, representing Christianity, went at his opponent with all the subtlety of a 'new atheist', mocking reincarnation and the Buddhist doctrine that there is no such thing as the self. The Buddhist, Gunananda, gave as good as he had got. He taunted the Christian for being unable to pronounce the sutras correctly, let alone understand them. As for the Bible, how come the Israelites needed to daub their doors with lambs' blood to identify themselves at the first Passover in Egypt? If God was really omniscient, he'd surely already know where they lived.

On the second day of the debate, da Silva moved on to cosmology. He noted that Mount Meru, the square massif at the centre of the Buddhist universe, was supposedly over a million miles high and wide. So, he asked, why was something so big invisible to the eyes of men? 'Climb to the top of the tall tree described in your sutras and you will definitely see it,' he sneered. Then he pointed to a globe that he'd placed on top of an inkstand. 'Now the circumference of the Earth is 25,000 miles,' he said. 'This is admitted by all the civilized nations of the world. This fact is proved by every day's experience.' There wasn't room for Mount Meru on Earth and if it stood in space, we'd all see it in the sky.[14]

'The Buddha' is an honorific applied to Prince Siddhartha Gautama (d. c. 400 BC). Born into luxury, he rebelled against the caste-based culture of his time and rejected the sacrifices to the gods stipulated by the Vedas that were the foundation of Brahminic power. Instead, he urged his followers to seek *nirvana* and preached an ethical middle way between asceticism and excess. The Maurya emperor Ashoka was an early convert to Buddhism after he was repelled by the bloodshed in a war of conquest that he had instigated himself. Ashoka's patronage helped spread the Buddha's teaching across India and beyond. An early tradition credits him with sending missionaries to Sri Lanka, where Buddhists still make up the majority of the population.[15]

In the fourth century AD, the monk Vasubandhu set out what has become the canonical description of the universe for South Asian Buddhists. Because it was rooted in the same Indo-European milieu, it shared some of the features with the world picture of the Puranas. He placed Mount Meru at the centre of a flat earth, where it rose to touch the heavens. The mountain was surrounded by an ocean, contained by an iron wall at the edge of the world. Four continents lay in the waters, one at each of the cardinal points around Meru. India was in the southern landmass, and journeys to the other continents were confined to the realm of myth. As in the Puranas, tiers of heavens and hells were stacked above and below the Earth.[16] Beautiful representations of this world picture survive in collections of Indian manuscripts.

As we've seen, one of the pressures that led to Indian astronomers adopting the theory of the Globe was the need for accurate timekeeping so that Vedic rites could be performed at the appropriate moment. The Buddha himself had explicitly rejected the need for these rituals, so his followers lacked one motivation for a more accurate calendar. Furthermore, Buddhist monks were specifically admonished from studying the so-called mundane sciences, such as astronomy, although this was not always observed. Likewise, the Buddha was said to have warned against astrology, closing off another incentive for observing the heavens.[17] There were gifted Buddhist astronomers, such as Yixing in China, but their studies of the heavens were largely divorced from their religious beliefs.

The Globe was a problem for Buddhists, however they sliced it. The issue was particularly acute in Japan after the Meiji Restoration overthrew the Shogunate in 1868. As society rapidly modernized, maps of the world became a familiar sight in schools. Japanese Buddhists had already been debating how to accommodate Western cosmology for over a century, without resolving the issue. A brilliant young scholar, Tominaga Nakamoto (1715–1746), developed a critical method of reading the sutras to distinguish between the original words of the Buddha and those that reflected the agendas of his followers. He found contradictory descriptions of Mount Meru, especially regarding its size. They couldn't all be right. Besides, he said, the Buddha's religious teaching didn't depend on Mount Meru, which simply reflected the Vedic environment in which he lived. 'This is because the Buddha's intention is not to be found in such matters,' Tominaga explained. 'He was urgently seeking people's salvation and had no time for such petty matters.'[18]

By the late nineteenth century, Japanese writers had set-tled on the idea that the world picture of the sutras was a later accretion to the message of the Buddha, or else an example of him speaking in the language of the ordinary people he was preaching to. If the Buddha returned today, they said, he'd have been perfectly happy to describe the Earth as a sphere. In the 1870s, teaching about Mount Meru was even made illegal in Japan.[19] Other Buddhists sought to describe it as a metaphor for modern cosmology. In 1927, a Chinese monk called Taixu (1890–1947) harmonized European and Buddhist world views by showing how elements of the latter could be mapped onto the solar system.[20]

Tibet was the last bastion of traditional Buddhist cos-mology. The fourteenth Dalai Lama (b. 1935) recalled that he had been taught the world picture of Vasubandhu, described above, while he was at school. But by the mid-1950s he had sufficient exposure to Western magazines to know that the Earth was round. The traditional model did not much appeal to him, and he later took the drastic step of declaring it wrong and contrary to reason.[21]

British missionaries to Sri Lanka had been quick to exploit the inconsistency between Western and Buddhist world pictures. Mount Meru could not exist, they said, because explorers had already visited enough of the planet to be certain there was nowhere it could be hiding. The evi-dence 'would all tend to prove that the Buddha was ignorant of the true figure of the Earth, and that all he says about it is unscientific and false'. Furthermore, the Earth must be round because a ship sailing west from Sri Lanka won't hit the bar-rier that rings the world, but would eventually end up back where it started. And if the Buddha was wrong about this, he was probably wrong about much else besides: 'he cannot, therefore, be a safe teacher.'[22]

So, how did the Buddhist monk, Gunananda, respond to David da Silva near Colombo in 1873? While European missionaries had introduced the Globe to the East, da Silva was a native Sinhalese convert to Christianity. The debate was between Sri Lankans of different faiths. It turned out that Gunananda had been doing his own research on Western science and had found that some people in England spurned the discoveries of Newton. If there were Europeans who rejected the axioms of modern science, he asked, why should Sri Lankans abandon their traditions? He had a point. Even England still had its flat-earthers.

Flat-Earth Societies

In 1849, Samuel Birley Rowbotham (1816–1884), under the pseudonym of Parallax, published a short pamphlet setting out why he thought the Earth was flat. Rowbotham, who had started his career at a socialist commune in East Anglia, promoted his work by travelling the country to give talks in village halls and at working men's clubs. He was an eloquent speaker and developed rejoinders to all the arguments for the Globe that his audience could throw at him. A frequent challenge was how it could be possible to circumnavigate the Earth if it was a disc. Rowbotham would respond that, since the North Pole is at the centre of the world, a ship sailing on a bearing perpendicular to due north would inevitably perform a circuit and end up back where it started. Some of the reviews of his performances by local journalists read like he had half-convinced them. He collated the arguments honed during his debates into expanded versions of his pamphlet: the final edition was a book of over four hundred pages, full of experiments and ripostes to prove the Earth was like a plate. By the time he published this definitive version of his arguments, Rowbotham had diversified into quack medicine.[23]

He was successful enough to be living comfortably in London with a young wife and a multitude of children.

To describe his flat-earth programme, Rowbotham coined the name 'zetetic astronomy', from the Greek word *zeteo* meaning 'seek', as in 'seek and you shall find' from Matthew's gospel.[24] The zetetic shtick majored on common sense. Anyone with eyes in their head could see for themselves the evidence that the Earth is flat. Rowbotham urged people to carry out experiments, while explaining what the results would be for those without the inclination to perform them for themselves. Zetetic astronomy was 'in every respect demonstrable philosophy, agreeing with the evidence of our senses, borne out by every fairly instituted experiment'. By contrast, 'The modern or Newtonian astronomy has none of these characteristics. The whole system taken together constitutes a most monstrous absurdity.'[25] Rowbotham insisted he would meet all challengers, but usually ducked and weaved to avoid a verifiable test of his theories. Once, cornered in Plymouth, he simply denied the fact that a lighthouse, some 40 kilometres (24 mi.) distant, was partially obscured by the curvature of the Earth. His chutzpah was so imposing that some observers went away convinced his utter failure to verify his theories had been a great success.

For all their insistence on common sense, religion was the primary motivation of the zetetics. They read the Bible literally, sprinkling their arguments with proof tests from scripture. As for modern scientists, 'Because the Newtonian theory is held to be true they are led to reject the Scriptures altogether, to ignore the worship, and doubt and deny the existence of a Creator and Supreme Ruler of the world.'[26] From time to time, there were lively debates in the pages of working men's periodicals such as the *English Mechanic*. The readership of these journals wanted to educate themselves while harbouring some distrust of the ruling class. They also tended to consider themselves good Christians.

These were the people the zetetics sought to convince, admittedly with a thoroughgoing lack of success. The vast majority of Britons identified as Christians, and almost none of them took the flat earth seriously.

One zetetic tactic was to bring notable scientists into the discussion. The Astronomer Royal George Airy (1801–1892) was plagued by letters from a few flat-earthers demanding he pay attention to their arguments. He found it astonishing that there were people who took this sort of thing seriously, so filed their letters under 'My Asylum for Lunatics' and quickly learned not to respond.[27] One of Rowbotham's most garrulous and aggressive followers, John Hampden (1819–1891), hooked a bigger fish by posting a wager of £500 that he could prove the Earth is flat. Alfred Russel Wallace, the co-discoverer of the theory of natural selection with Charles Darwin, picked up the gauntlet. Unlike Darwin, Wallace was not privately wealthy and, at the time, salaried jobs in science were in short supply. Hampden's offer of £500 to prove the Earth isn't flat seemed like easy money.

Wallace and Hampden set the challenge for 28 February 1870 at Welney Bridge over the Old Bedford River, a seventeenth-century drainage canal in the Norfolk fens that ran straight and level for many miles. Wallace and Hampden agreed they would settle their bet on a 10-kilometre (6 mi.) stretch near the town of Wisbech. This had been Rowbotham's old stomping ground, where he'd carried out his observations to prove, at least to his satisfaction, that the Earth is a plane. On the appointed day, Wallace set up posts along the river, with a scale on them. He calculated that for each mile distant, 7.6 centimetres (3 in.) of the posts should be obscured by the bulge of the Earth. It turned out that the posts were too far away to read the scale, but a few days later, armed with a powerful telescope, he convinced himself and a second neutral observer that this was precisely what he saw. Hampden obviously denied it.

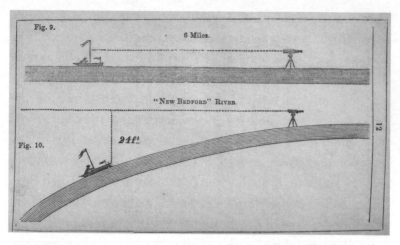

Rowbotham's diagram of Bedford Bridge, from Samuel Rowbotham ('Parallax'),
Zetetic Astronomy (1849).

Unfortunately, he had deceived Wallace into thinking that the other third-party observer was also impartial, whereas in fact he was a fellow zetetic. With the flat-earthers denying the evidence of their eyes, Wallace was at an impasse. For the bet to be settled, both observers had to be convinced.[28]

Once Hampden's cheating was exposed, Wallace received the £500. This merely exacerbated his troubles. Hampden pursued him through the courts and showered him with all sorts of calumnies. When he started sending postcards to Wallace's wife as well, the poor man sued. Hampden was hauled before the Lord Chief Justice and found guilty of libel. When he couldn't afford the damages, he was thrown into prison.[29] None of this really helped Wallace. Another court found the wager was invalid and he had to return the money. The experience left him out of pocket and regretful of ever getting involved.[30]

After the deaths of Rowbotham and Hampden, the torch of the flat-earth movement passed to an indomitable woman called Lady Elizabeth Blount (1850–1935). She was a campaigner for many quirky causes that were deeply unfashionable in her day.

As well as being a vegetarian, she advocated for animal rights and against vivisection, not to mention challenging contemporary views about the superiority of white people over other races.[31] Her marriage to a baronet gained her a title and access to money, which she used to promote her views through journals and public speaking. In 1904, she and her husband arranged for a repeat of the Bedford River experiment, this time using a camera with a telephoto lens. The photographer took a picture that purported to show a white sheet near the surface of the river at a distance of 10 kilometres (6 mi.).[32] It should have been below the horizon but may well have become visible due to atmospheric refraction. In any case, Lady Elizabeth triumphantly distributed the picture far and wide as proof that the Earth is flat. After her death in 1935, English zeteticism flagged without her energy to sustain it. But it would soon enjoy a revival on both sides of the Atlantic that has lasted to the present day.

23

Today: 'There's nothing particularly exciting about a round world'

One of the few modern Western societies in which belief in the flatness of the Earth was institutionalized was the city of Zion, a few miles north of Chicago on the shores of Lake Michigan in the United States. It was founded in 1901 to be a sanctuary organized along Pentecostal Christian lines. From the start, it was an unusual place, ruled over by charismatic religious leaders who stamped their credo onto the city's law book. Municipal rules banned diabolical evils like alcohol, gambling, prostitution and tobacco, not to mention less widely condemned vices such as opera, vaccination and modern medicine. Following the dietary rules in the Old Testament, the authorities forbade the consumption of pork and shellfish.[1]

Under the direction of Wilbur Glenn Voliva (1870–1942), Zion instilled its ethos into its citizens from a young age. Naturally, biology lessons echoed the Creation accounts in the Book of Genesis. In their geography classes, children in the local schools learnt that the Earth was a flat disc and that the Sun was just a few thousand kilometres distant. Voliva also promoted his world picture on the city's radio station, using a powerful transmitter that meant his preaching could be heard far from northern Illinois.[2]

For a while, the city prospered. Its industrial bakery produced the Zion Fig Bar, a bestselling snack until the 1950s.[3] It couldn't last. Like many Pentecostalists of the period, Voliva had convinced himself that the apocalypse was around the corner.

He devoted much of his time to spreading the word about the end times, while the city chafed under his authoritarian leadership and strict biblical injunctions. By 1935, residents had had enough. They deposed Voliva, who retired to Florida, and set about normalizing Zion's constitution along conventional lines.[4] From then on, the task of keeping the flat-earth flame alive fell to a valiant few.

The International Flat Earth Research Society

In 1956 Samuel Shenton (1903–1971), a sign-maker from southern England, founded the International Flat Earth Research Society (IFERS). Like the zetetics of the Victorian era, Shenton was a conservative Christian, who based his quixotic campaign on empiricism and common sense.

He couldn't have picked a worse time to launch IFERS. The following year, an unprecedented challenge to the flat earth emerged from the Soviet Union. On 4 October 1957, Sputnik, a tiny satellite the size of a beach ball, orbited Earth, emitting a radio signal. This looked like undeniable evidence that our planet was round. As the space race progressed over the next decade, direct evidence for the Globe piled up. Manned missions, pictures of Earth from space and close-ups of the Moon threatened to puncture even Shenton's confidence about the two-dimensionality of the world. His pronouncements became increasingly hyperbolic and religiously grounded. Then, on 23 December 1968, came the *coup de grâce*. The Apollo 8 mission carried three crew into orbit around the Moon. In a live broadcast from their capsule, albeit in grainy black and white, the astronauts turned the camera towards the window where Earth was clearly visible. The next day, they added insult to injury. In a Christmas message to the people of the world, they quoted the first lines of Genesis: 'In the beginning, God created the heaven

'Earthrise' image taken by astronaut Bill Anders while in lunar orbit
during the Apollo 8 mission, 1968.

and the Earth . . .'[5] The astronaut Bill Anders (b. 1933) took a
glorious photograph of the Earth, half in darkness, rising over
the surface of the Moon.

Against this background of American achievement, English
journalists needed a local angle. Newspapers, always on the
lookout for eccentric characters, alighted on Shenton's IFERS
and turned him into a minor celebrity. He exhausted himself
giving interviews and replying to correspondence defending the
flat earth, all the while desperate to be taken seriously by the
reporters who were laughing at him.

After Shenton's death, his campaign was taken up by a
Californian called Charles Johnson (1924–2001). From the

1960s, fundamentalist Christian groups had promoted 'young-earth creationism', rejecting evolution and insisting the Earth was created in six days in the last few thousand years. Like the zetetics, these creationists dressed their religious dogmas in scientific clothes. However, even they wouldn't touch flat-earth theory, to the great disappointment of Johnson and his wife, Marjory. The couple felt they were the last people still true to the teachings of the Bible.

Marjory had emigrated from Australia to the United States in 1944, and when she arrived was outraged to find her native country referred to as 'Down Under'. She swore she had never hung upside down and that almost all Australians were flat-earthers.[6] Meanwhile, Johnson railed against 'greaseball science' and developed an elaborate alternative history of the world that explained how globalism had triumphed. Alas, all their efforts met with a singular lack of success. When their house burned down in 1995, the flames consumed most of the archives of the IFERS. It felt like the end.[7]

And yet, today, flat-earthism is in rude health. It turned out its proponents were not as recherché as they assumed, although it took the Internet to bring them together. Videos on YouTube promote their views. Google has attempted to curtail conspiracy theories on its platform, but counter-productions by sceptics only serve to spread the message more widely. Nowadays, conventions of flat-earthers can attract hundreds of people. What unites them is a distrust of authority and government. They are likely to have encountered the idea after investigating whether the Moon landings were faked or 9/11 was an inside job.[8]

Belief in a flat earth is the uber-conspiracy theory: it requires its adherents to think absolutely everyone else is deliberately deceiving them and that the entire infrastructure of modern life is fraudulent. For that reason, flat-earthers are an object of scorn, if not pity. However, unlike anti-vaxxers, to date, they have harmed only themselves. In February 2020, a dare-devil called 'Mad' Mike Hughes was killed in his homemade steam-powered rocket while allegedly endeavouring to prove that the Earth is flat.[9] Perhaps we shouldn't mock – he had, at least, the courage of his convictions.

The Great Conflict between Science and Religion

Today, the flat earth is indelibly associated with foolishness and ignorance. That's reflected in popular perceptions of history: the ancient Greeks were smart and knew the world is round, but people in the Middle Ages were under the thumb of a dog-matic church, so they must have been flat-earthers, unlike the wise Muslim scientists of Baghdad. As we've seen, the truth isn't so clear-cut. The theory of the Globe was well established in medieval Christendom, but was only accepted slowly in many other parts of the world.

The myth that everyone thought the Earth was flat until Columbus proved otherwise was taught as a fact from nurs-ery to university for much of the twentieth century. Textbooks regurgitated the falsehood until it became an unexamined part of conventional wisdom.[10] It wasn't just the general public who believed the fiction of medieval stupidity and supersti-tion. Notable historians of the late twentieth century peddled the calumny too. J. H. Parry (1914–1982), a British academic who became a Harvard professor, with a list of honours as long as his arm, wrote that Cosmas Indicopleustes 'had exerted a formidable, though never quite unchallenged, influence upon cosmographical thought' in western Europe through to the late Middle Ages.[11] In fact, it is unlikely that anybody in the West had even heard of Cosmas until a Latin translation of his book appeared in the eighteenth century. Daniel Boorstin (1914–2004), the Pulitzer Prize-winning Librarian of Congress, devoted an entire chapter of his history of science, *The Discoverers*, to the belief in a flat earth during the Middle Ages.[12]

Some scholars knew better. In a thin volume from 1991, Jeffrey Burton Russell (b. 1934) launched a head-on attack against the mythical medieval belief in a flat earth, although, unfortunately, he ended up understating the extent to which the tabernacle of the Antioch world picture had influenced early Christians.[13] The Italian cultural historian Umberto Eco (1932–2016), famous for his novel *The Name of the Rose*, also fought back against the error. In a talk at the University of Bologna in 1994, he explained that Columbus's opponents at the Council of Salamanca certainly didn't think the Earth was flat. Unfortunately, the memo never reached his British publishers. The blurb on the back of the book containing the English translation of the lecture said that believers in a flat earth had accidentally encouraged Christopher Columbus to make his voyage across the Atlantic.[14]

The false belief in a medieval flat earth is just one element in the larger fallacy that science and religion have been locked in a great conflict. The origins of this thesis can be found in the Reformation, when half of Europe rejected the authority of the Catholic Church. To justify the break with Rome, Protestant propaganda portrayed the Church in the Middle Ages as a font of ignorance and superstition. For example, Sir Francis Bacon (1561–1626), seeking to justify his belief that Catholicism had held back science, wrote, 'On the accusation of some of the early Church Fathers, Cosmographers, who on clear evidence that no sane man would reject today, claimed the Earth was a sphere and therefore inhabited at the antipodes . . . were brought to trial for impiety.'[15] This looks like a garbled recollection of the accusation against Virgil of Salzburg in the early Middle Ages.

The myth was taken up by Thomas Paine (1737–1809) in the eighteenth century, who said that poor Virgil was 'condemned to be burned for asserting the antipodes, or in other words, that the Earth is a globe'.[16] One would think that having 'burnt as a heretic' on his CV would have complicated Virgil's successful bid for canonization. The American founding father Thomas Jefferson (1743–1826) managed to get things even more wrong. In his *Notes on the State of Virginia*, he said that the Inquisition prosecuted Galileo for teaching that the Earth was spherical.[17]

After the French Revolution, the gloves came off. Respected scholars, such as Antoine-Jean Letronne (1787–1848), a professor of history at the Collège de France, transformed the myth into an academic commonplace.[18] This became part of an overarching thesis that science and religion were locked in an eternal struggle, a fight that science was winning. In England, polemicists, such as 'Darwin's bulldog' Thomas Huxley (1825–1895) and the physicist John Tyndall (1820–1893), promoted this view. Later in the century, two famous books cemented the conflict between science and religion into the public consciousness. These were the *History of the Conflict between Religion and Science* by John William Draper (1811–1882) and the *History of the Warfare of Science with Theology in Christendom* by Andrew Dickson White (1832–1918). White's take on Columbus was more colourful than most:

Many a bold navigator, who was quite ready to brave pirates and tempests, trembled at the thought of tumbling with his ship into one of the openings into hell which a widespread belief placed in the Atlantic at some unknown distance from Europe. This terror among sailors was one of the main obstacles in the great voyage of Columbus.[19]

More recently, the conflict thesis made it into the TV show *The Simpsons*, in an episode with a guest voice from the palaeontologist Stephen J. Gould (1941–2002).[20] Gould himself would go on to write *Rocks of Ages*, a book-length rebuttal of the conflict thesis, and the medieval flat-earth myth in particular.[21]

Nowadays, historians spurn one-dimensional accounts of the relationship between faith and science. They've recognized that much of the momentum behind the conflict thesis was the anti-Catholic bias pervasive in the nineteenth century. As for Draper and White, while neither man was an atheist, they were as resolute against the papacy as the next English-speaking liberal. Indeed, their distaste for religion didn't extend much further than Rome. Draper's book was so sympathetic towards Islam that it was translated into Turkish as part of the movement to modernize the Ottoman Empire.[22] Both men were seeking to reconcile a reformed religion to science but accidentally further estranged them.[23]

This book might seem to provide grist for the mill of the conflict thesis. After all, we have seen how much of the resistance to the Globe came in clerical garb. For instance, the zetetics saw their atavistic campaign as part of an irreconcilable struggle between modern science and Christianity. But religious figures were prominent on the other side of the debate as well. Isidore of Seville and the Venerable Bede might have disagreed about the shape of the Earth, but they are both honoured as saints by the Catholic Church. Muslim scholars long differed on the shape of the Earth, but both sides of the dispute were as devout as each other. Indian astronomers first justified the Globe as a way to improve the accuracy of the calendar, specifically so that Vedic rituals would be performed at the right moment. It is debatable whether Confucianism is a religion, but much of the learned elite of China still held on to their traditional world picture even after the arrival of the Jesuits. As for Lucretius, he is today hailed as a champion of rationalism, to the extent that it's largely been forgotten that he thought the Earth was flat.

Ultimately, though, the alleged conflict between science and religion isn't about great themes like 'faith and reason' or 'free thought versus conformity'. It is simply a question of chronology. Sacred texts like the Bible and Koran are written in everyday language and, aside from their core religious messages, reflect the conventional wisdom of their original audience. So, in as much as they include any comment about the shape of the Earth, or other scientific subjects, it is just whatever stood for general knowledge at the time they were composed. When new ideas come along, there is bound to be a disagreement with older texts, just as the likes of Galileo rendered much of Aristotle's physics obsolete. Most of the time, contradictions between scripture and scientific developments get resolved without too much fuss. Admittedly, the Catholic Church foolishly nailed its colours to Aristotle's mast and, today, creationists still challenge Darwinism because it conflicts with the Bible. However, these disputes have been exceptions in a relationship between science and religion that has generally been harmonious.

As for the flat earth, it lives on in the imaginations of some of our finest writers.

Fantasy Worlds

In January 1926, a new fellow of Magdalen College, Oxford, gave his first lecture. It was standing room only, and he had to lead his audience into a larger room, so they could all sit down.[24] This was the start of a career in which he educated generations of undergraduates. One course of his lectures, delivered many times at Oxford, was eventually published as *The Discarded Image: An Introduction to Medieval and Renaissance Literature* in 1964. This don's name was C. S. Lewis (1898–1963), who had died the previous year, celebrated for his Narnia series of children's books, as a Christian apologist and an expert on late medieval English literature. In his academic work, Lewis fought back against the myth that medieval people thought the Earth was flat. As he wrote in *The Discarded Image*, 'Physically considered, the Earth is a globe; all the authors of the high Middle Ages are agreed on this . . . The implications of a spherical Earth were fully grasped.'[25]

However, when it came to creating his own world, Lewis found the Globe was not fit for his purposes. Narnia was flat. In the third Narnia novel, *The Voyage of the Dawn Treader*, King Caspian leads an expedition across the Eastern Ocean to the end of the world. As the ship sails towards the dawn, the rising sun grows progressively larger, eventually becoming five or six times bigger. On the far side of the Ocean, the water is fresh and shallow, and covered with lilies. When the draught of the ship means it can go no further, a mouse, who is part of the crew, boards a tiny rowboat and paddles his way through a wall of water where he disappears. Beyond, the people on deck can just make out the high mountains of Aslan's land, perhaps the earthly paradise of medieval legend.[26]

The amount of commentary on C. S. Lewis, by both scholars and fans, is vast. Many have written about Narnia's geography, but no one has given an adequate solution to why Lewis made it flat. After all, he was at pains to note in *The Discarded Image* that this was not how medieval people saw their world. Rather than over-analysing the issue, it seems likely we can find the answer in an exchange between King Caspian and a pair of English schoolboys called Edmund and Eustace, who have been magically transported to the *Dawn Treader*. Eustace thinks it ridiculous to say the world has an edge, but Edmund realizes that Narnia might be different to our own planet. Caspian is thrilled to learn about this:

'Do you mean to say,' asked Caspian, 'that you
three come from a round world (round like a ball)
and you've never told me! It's really too bad of
you. Because we have fairy tales in which there
are round worlds, and I've always loved them.
I never believed that there were any real ones . . .
It must be exciting to live on a thing like a ball.
Have you ever been to the parts where people
walk about upside down?'
  Edmund shook his head. 'And it isn't like
that,' he added. 'There's nothing particularly
exciting about a round world when you are
there.'[27]

Lewis seemed to grasp how fantastical the theory of the
Globe would be for someone who lived on a flat world – more
like a myth than reality. But he also saw (this is a point he made
many times) how inured to wonder we become once we take
something for granted. Narnia was flat because it's modelled
on the *mappa mundi*, rich in Christian symbolism and different
from the world revealed by science. But Lewis also wanted us
to understand that it is possible to move between the symbolic
and the real, if you know how. And from the inside, the symbolic world looks just as real as ours.

C. S. Lewis was not the only author to make his fantasy
world flat. His friend J.R.R. Tolkien (1892–1973), another
Oxford expert on medieval literature, did the same. The events
in *The Lord of the Rings* and *The Hobbit* take place on a spherical earth. However, this had not always been the shape of the
world. Thousands of years prior to the events in those books,
evil men from the island of Númenor had attempted to invade
the land of the elves in the far west. The dark lord Sauron had
corrupted Númenóreans into believing that by doing so, they

could obtain immortality. The elves prayed for deliverance. God, called Ilúvatar, uprooted the Undying Lands from the circle of the world, making it completely inaccessible to men. A great chasm opened up, which swallowed Númenor. Finally, the gods, authorized by Ilúvatar, 'bent back the edges of Middle Earth, and . . . they made it into a globe, so that however far a man should sail he could never again reach the true West, but came back weary at last to the place of his beginning'.[28]

Tolkien's academic speciality was Anglo-Saxon literature, and his particular passion was the Old English epic *Beowulf.* This story is set in a world that is a disc. In 'The Monster and the Critics', a lecture Tolkien gave on *Beowulf* in 1936, he noted that the poet 'and his hearers were thinking of the *eormengrund*, the great Earth, ringed with *garsecg*, the shoreless sea, beneath the sky's inaccessible roof'.[29] He expanded on this point in his commentary on the poem, explaining that while educated people around AD 800 might well have been aware the Earth is round, this is not reflected in the verse and probably wasn't part of the imaginative world of its hearers either.[30]

England doesn't have a fully developed mythology like the Greeks or Vikings. Perhaps it was lost or it never existed. Tolkien set out to create one. As this wasn't a scholarly exercise, he was free to pilfer tropes from surviving stories in other related cultures, be they Irish sagas or French romances. For example, at the dawn of time, Tolkien wrote, the whole land was illuminated by two trees – an idea inspired by a passage from one of the medieval romances about Alexander the Great.[31] When these trees were destroyed, fragments of their light formed the Sun and the Moon.

According to Norse mythology – another influence on Tolkien – Yggdrasil, a great ash tree, stood at the centre of the world. In its roots, the worlds of men and giants were to be found. Even though they were great seafarers who briefly colonized North America, there was no room for the Globe in the Viking imagination. As late as the thirteenth century, the Icelandic chronicler Snorri Sturluson (1178–1241) began a history with the Old Norse words meaning the 'the world disc', which gives the saga its modern name *Heimskringla*.[32]

The reason Tolkien made Middle Earth flat in its earliest incarnation may have been recognition that this is how all peoples once saw their world. The sundering of the Undying Lands and Middle Earth represented a loss of innocence that shut humanity off from the spiritual realm. Yet Tolkien came to regret this aspect of his mythos. In a lecture in 1939, 'On Fairy-Stories', Tolkien had discussed the importance of 'secondary belief' and the rules that fantasy novels had to follow to achieve verisimilitude.[33] He came to regard the flat earth as failing the test because it was 'astronomically absurd'. Unfortunately, it was too late to change such an integral part of the story of the Silmarillion, which might be why he never completed it.[34] C. S. Lewis revelled in including absurdities in his tales (such as a lamp post in Narnia and an appearance by Father Christmas). Tolkien could not abide them.

Lewis knew that the absurd can teach us more about the human condition than any amount of consistency. Of the great British fantasy authors, it was Sir Terry Pratchett (1948–2015) who took this as his lodestone. All the basic elements of Pratchett's world were present in the first Discworld novel, *The Colour of Magic*, published in 1983. As explained in the book's solitary and lengthy endnote, Discworld is, axiomatically, a disc supported by four elephants, which stand on the back of the cosmic turtle A'Tuin. The sex of the turtle is unknown, but in the second novel, *The Light Fantastic*, it supervises a cache of eggs as they hatch to reveal baby cosmic turtles, each with a coterie of elephants. As we've seen, the Indian astronomer Lalla had been complaining about people who thought the Earth sat on the back of a turtle or pachyderm way back in the eighth century. Pratchett's excuse was that his world had to be flat because it runs on a mixture of magic and narrative. The momentum of the story meant that his protagonists would end up falling off. As the hapless wizard Rincewind said, when caught in the current caused by the Rimfall, 'We're being carried over the edge of the world.'[35]

Pratchett never let the message get in the way of the medium. He wasn't attempting to dazzle his readers with the astonishing imaginary world he had created. He was teaching them what it meant to be a human being. And that was subsidiary to the imperative of keeping them entertained. Since he wore his genius lightly, Pratchett has not yet garnered the scholarly attention he warrants. Perhaps, when he does, we will understand the sources he used for his work and discover exactly how he landed on the concept of the Discworld.[36]

C. S. Lewis, J.R.R. Tolkien and Terry Pratchett believed that a flat world showed us things about our own that we might not otherwise notice. They also recognized that the Globe is startling, counter-intuitive and defies common sense. It's not surprising it took so long for everyone to accept it.

Afterword

Everywhere we look, there are globes. Between 1963 and 2002, the ident used by BBC1 between television programmes was a globe. People in the UK saw it several times a day. Globes are still popular symbols with broadcasters in much of the world, especially for news bulletins. Travel firms like to incorporate one into their branding too. The International Air Transport Association and the Federal Aviation Authority both have one. The United Nations, on the other hand, uses the flat earth for its logo, which has given succour to conspiracy theorists.

The ubiquity of globes has normalized the idea that our planet is a sphere. Most of us were so young when we first learnt about it that we cannot even remember the occasion. We rarely think about what a strange thing it is or how hard it must have been to accept in the pre-modern world. Throughout this book, I've maintained that it is common sense to think the Earth is flat. All cultures once believed this, even if their world pictures were otherwise influenced by their environment and society.

When a scientific theory seeks to overturn our everyday experience, whether the theory claims that the Earth rotates, or species change or time dilates, it also needs to show why we perceive things the way we do. It's not enough to claim that we are orbiting the Sun if you don't also explain why we cannot feel that we are moving.[1] This meant that, to produce a convincing argument for the Globe, Aristotle had to invent a geocentric universe in which heavy objects were attracted to the centre. Only by redefining what he meant by downwards and showing this new definition was consistent with our instincts could he hope to carry the day. The empirical evidence of the horizon and the visibility of stars was not enough alone to persuade people that the Earth is round. The Chinese knew these things but didn't deduce our planet's shape as a result. However, by marrying observation to a new world picture, Aristotle was able to convince people of his ideas. Astronomers, in particular, cemented acceptance of the Globe by using it as a foundation for impressively accurate models of the universe.

Educated Christians in the Greek-speaking world generally accepted Aristotle's cosmology because it was the milieu in which they had been brought up. But it was rejected by the influential Antioch School, together with people who lacked education in the liberal arts. They held to alternative world pictures, such as the tabernacle model, which survived for centuries in the Byzantine Empire and the Church of the East.

Following the fall of the western Roman Empire, Latin-speaking Christians lost access to Greek learning. This initially led to a sense of disorientation about the shape of the world, as we saw in the works of Isidore of Seville. However, renewed attention to classical authorities like Pliny convinced the Venerable Bede that the theory of the Globe had to be accepted. Early medieval Christians suffered from a cultural inferiority complex that led them to assume ancient writers were right as long as what they said could be reconciled to the Bible. Thankfully, while both the Old and New Testaments assume the Earth is flat, they are sufficiently vague on the issue to be interpreted in other ways as necessary.

The prestige of Greek science was the major reason that Indian astronomers adopted the Aristotelian world picture as they sought to improve their calendar and the almanacs used for astrology. Likewise, the esteem in which Indian wisdom was held in Persia and China led to books containing the theory of the Globe being imported to both nations. This alone was not enough to overcome the traditional Chinese world picture of a square earth and circular heaven. However, among the Muslim philosophers of the Abbasid caliphate, the spherical earth was widely adopted. They saw themselves as the rightful inheritors of Greek scientific knowledge and improved the astronomical models of Ptolemy. Patronage from the caliph enabled them to carry out the long-distance observations necessary to validate the Globe. It was also helpful when ensuring mosques were correctly aligned towards Mecca. Nonetheless, traditional world pictures, buttressed by sacred writings and common sense, coexisted with the Globe in India and the Islamic world until modern times.

In some cultures, a spherical earth faced particular challenges. For Manichaeans and Zoroastrians, the shape of the world was of critical importance. The struggle between light and dark was fought between forces that emanated from above and below. It proved too hard to adapt theological history to fit within a spherical universe, although Dante showed in his *Divine Comedy* that it would have been possible if it was absolutely necessary. The world picture of ancient China made adoption of the Globe difficult as well. Chinese bureaucrats, the mandarins steeped in the Confucian Classics, took the correlations between the natural and human domains extremely seriously. Their model of the universe was built into their ways of thinking and even featured in town planning. Meanwhile, Chinese astronomers had developed sophisticated methods that kept their calendars more accurate than the Western one, at least until the sixteenth century. The emperor's advisers could accept employing foreign personnel in the astronomical bureau but saw no reason to employ foreign cosmology. They were confident in China's ancient, tested and consistent world picture. China eventually succumbed to the Globe during the colonial era. Western prowess in science and technology was impossible to deny, although the bitterness of this pill was alleviated by claiming European astronomy had Chinese origins.

As the rest of the world drew together into a sphere in the nineteenth century, there were some who looked back to simpler times, when the Bible could be taken to contain a full description of nature. Today, the various flat-earth societies join a plethora of conspiracy theories that have found a home on the Internet. Perhaps they serve as a salutary reminder that our modern world picture is not obvious and we shouldn't be too hard on those who still won't swallow it.

The Earth is spherical. That statement is true, not because we believe it but because it corresponds to reality. The majority of Western philosophers would agree this is how to determine whether a proposition is correct. As we've seen, they even have a name for it: the correspondence theory of truth. But that is not how human beings think. When we hear a new piece of information, we are likely to concur only if it fits in snugly with everything else we know. In particular, we're more ready to believe something we learn from a source we trust. That's why people listen more attentively to Sir David Attenborough (b. 1926) on climate change than they do to politicians. In other words, our perception of truth is relative. Philosophers have a name for this too: the coherence theory of truth. We accede to a proposition if it is consistent with other propositions we accept as true. While the correspondence theory is surely correct, the coherence theory much better describes the process by which we come to believe something, whether or not that thing is correct.

The coherence theory of truth goes a long way towards explaining why the Globe was more readily accepted in some places than others. For example, when it came to cosmology, the Venerable Bede trusted the ancients more than he did the befuddlement of Isidore of Seville. He had never seen the star Canopus, but he knew it existed because Pliny said it did. By contrast, the Muslim philosopher Averroes (1126–1198) was able to observe for himself that Canopus was visible in Marrakesh but not in his native Cordoba.[2] For some other Muslims, the Globe didn't agree with the plain meaning of the Koran, so it took longer for the idea to gain acceptance. There was a similar process at work in India – whether someone believed in the Globe depended on the degree to which they saw the Greeks as more reliable purveyors of cosmology than the Puranas.

This might also explain the longevity of the flat earth in Chinese thought. It cohered with the Confucian world view of correlative cosmology. It was also a feature of the oldest and most revered texts, which members of the Chinese elite imbibed at an early age. Strangers from an inferior and younger civilization might have said that the Earth was a sphere, but they were not qualified to overthrow the wisdom of old. So, even though the Chinese elite were aware of the Globe, they had good reasons to reject it.

As for us, we don't believe the planet is round because we're more clever or rational than the people who came before us; we're just lucky enough to live in a society that takes it for granted. Aristotle did not have that advantage in fourth-century BC Athens. He deserves tremendous credit for discovering the theory of the Globe. After all, it was never independently developed anywhere else. While very little of the scientific lore of ancient Greece is still considered pertinent today, the Globe has stood the test of time. That makes it humanity's first great scientific achievement.

REFERENCES

Introduction: 'The Blue Marble'

1 Eugene A. Cernan, 'Blue Marble – Image of the Earth from Apollo 17', www.nasa.gov, 30 November 2007.
2 Jennifer Epstein, 'Obama Hits GOP on Fuel Rhetoric', www.politico.com, 15 March 2012.

1 Babylon: 'The four quarters of the Earth'

1 Irving Finkel, *The Ark before Noah: Decoding the Story of the Flood* (London, 2014), p. 82.
2 Ibid., p. 37.
3 Ibid., p. 38.
4 Ibid., p. 29.
5 Ibid., p. 4.
6 N. K. Sanders, ed., *The Epic of Gilgamesh* (Harmondsworth, 1972), p. 108.
7 Georges Roux, *Ancient Iraq* (London, 1992), p. 201.
8 Ibid., p. 398.
9 'The Babylonian Creation', in *Poems of Heaven and Hell from Ancient Mesopotamia*, ed. N. K. Sanders (Harmondsworth, 1971), p. 92.
10 Wayne Horowitz, *Mesopotamian Cosmic Geography* (Winona Lake, IN, 1998), p. 318.
11 J. Edward Wright, *The Early History of Heaven* (New York, 2000), p. 34.
12 John H. Walton, *Ancient Near Eastern Thought and the Old Testament: Introducing the Conceptual World of the Hebrew Bible* (Grand Rapids, MI, 2018), p. 86.
13 Roux, *Ancient Iraq*, p. 361.
14 Alexander Jones, *A Portable Cosmos: Revealing the Antikythera Mechanism, Scientific Wonder of the Ancient World* (Oxford, 2017), p. 128.
15 Ulla Koch-Westenholz, 'Babylonian Views of Eclipses', *Res Orientalis*, XIII (2001), p. 74.

313

16 Roux, *Ancient Iraq*, p. 183.
17 Francesca Rochberg, *Before Nature: Cuneiform Knowledge and the History of Science* (Chicago, IL, 2016), p. 76.
18 Jones, *A Portable Cosmos*, p. 141.
19 Koch-Westenholz, 'Babylonian Views of Eclipses', p. 72.
20 Francesca Rochberg, *The Heavenly Writing: Divination, Horoscopy, and Astronomy in Mesopotamian Culture* (Cambridge, 2004), p. 278.
21 Rochberg, *Before Nature*, p. 263.

2 Egypt: 'The black loam and the red sand'

1 Herodotus, *The Histories*, trans. Aubrey de Sélincourt (Harmondsworth, 1996), p. 88 (2:5).
2 J. M. Plumley, 'The Cosmology of Ancient Egypt', in *Ancient Cosmologies*, ed. Carmen Blacker and Michael Loewe (London, 1975), p. 29.
3 Joyce Tyldesley, *The Penguin Book of Myths and Legends of Ancient Egypt* (London, 2011), p. 38.
4 Marshall Clagett, *Ancient Egyptian Science: A Source Book*, 3 vols (Philadelphia, PA, 1995), vol. II, p. 375.
5 Tyldesley, *Myths and Legends of Ancient Egypt*, p. 90.
6 Rolf Krauss, 'Egyptian Calendars and Astronomy', in *The Cambridge History of Science*, vol. I: *Ancient Science*, ed. Alexander Jones and Liba Taub (Cambridge, 2003), p. 132.
7 J. Edward Wright, *The Early History of Heaven* (New York, 2000), p. 15.
8 Plumley, 'The Cosmology of Ancient Egypt', p. 37.
9 Krauss, 'Egyptian Calendars and Astronomy', p. 143.
10 Otto Neugebauer, *The Exact Sciences in Antiquity* (New York, 1969), p. 91.

3 Persia: 'Order and deceit'

1 Barry Cunliffe, *By Steppe, Desert, and Ocean: The Birth of Eurasia* (Oxford, 2015), p. 136.
2 Touraj Daryaee, *Sasanian Persia: The Rise and Fall of an Empire* (London, 2009), p. 70.
3 M. L. West, *The Hymns of Zoroaster* (London, 2010), p. 3.
4 Prods Oktor Skjærvø, *The Spirit of Zoroastrianism* (New Haven, CT, 2011), p. 81.
5 Ibid., p. 83.
6 Ibid., p. 51.
7 Ibid., p. 228.
8 James B. Pritchard, *Ancient Near Eastern Texts Relating to the Old Testament* (Princeton, NJ, 1969), p. 315.
9 Francesca Rochberg-Halton, 'Babylonian Horoscopes and Their Sources', *Orientalia*, LVIII/1 (1989), p. 104.

10 Alexander Jones, *A Portable Cosmos: Revealing the Antikythera Mechanism, Scientific Wonder of the Ancient World* (Oxford, 2017), p. 104.

4 Archaic Greece: 'The Shield of Achilles'

1 Homer, *The Iliad*, ed. William F. Wyatt and A. T. Murray (Cambridge, MA, 1924), vol. II, p. 333 (18:607).
2 Ibid., p. 323 (18:484).
3 Ibid., vol. I, p. 387 (8:485).
4 Ibid., p. 351 (8:14).
5 Ibid., vol. II, p. 195 (16:426–57).
6 Francis Cornford, *From Religion to Philosophy: A Study in the Origins of Western Speculation* (New York, 1957), p. 55.
7 Hesiod, *Theogony*, ed. M. L. West (Oxford, 1966), p. 44.
8 Hesiod, 'Works and Days', in *Hesiod and Theognis*, trans. Dorothea Wender (Harmondsworth, 1973), p. 77 (566).
9 Hesiod, 'Theogony', in *Hesiod and Theognis*, ed. Wender, p. 46 (720).
10 James S. Romm, *The Edges of the Earth in Ancient Thought: Geography, Exploration, and Fiction* (Princeton, NJ, 1992), p. 14.
11 Plutarch, 'Concerning the Face Which Appears in the Orb of the Moon', in *Moralia*, trans. Harold Cherniss and W. C. Helmbold (Cambridge, MA, 1957), vol. XII, p. 59 (923C).

5 The Origins of Greek Thought: 'Equally distant from all extremes'

1 Bertrand Russell, *History of Western Philosophy and Its Connection with Political and Social Circumstances from the Earliest Times to the Present Day* (London, 1961), p. 25.
2 G.E.R. Lloyd, *Magic, Reason and Experience: Studies in the Origins and Development of Greek Science* (Cambridge, 1979), p. 251.
3 Thucydides, *History of the Peloponnesian War*, trans. Rex Warner (Harmondsworth, 1972), p. 223 (3:49).
4 Jean-Pierre Vernant, *The Origins of Greek Thought* (Ithaca, NY, 1982), p. 62.
5 Pseudo-Xenophon, 'The Constitution of Athens', in *Scripta Minora*, trans. E. C. Marchant and G. W. Bowersock (Cambridge, MA, 1925), p. 475 (1.2).
6 Lloyd, *Magic, Reason and Experience*, p. 266.
7 Peter Harrison, *The Territories of Science and Religion* (Chicago, IL, 2015), p. 26.
8 Plato, *Theaetetus*, trans. Robin A. H. Waterfield (Harmondsworth, 1987), p. 69 (174a).
9 Daniel W. Graham, *Science before Socrates: Parmenides, Anaxagoras, and the New Astronomy* (Oxford, 2013), p. 51.
10 G. S. Kirk, J. E. Raven and M. Schofield, *The Presocratic Philosophers: A Critical History with a Selection of Texts* (Cambridge, 1983), p. 89.

11 Ibid., p. 107.
12 Ibid., p. 158.
13 Aristotle, 'On the Heavens', in *The Complete Works of Aristotle*, ed. Jonathan Barnes (Princeton, NJ, 1984), p. 484 (294a).
14 Kirk, Raven and Schofield, *The Presocratic Philosophers*, p. 133.
15 Aristotle, 'On the Heavens', p. 486 (295b).
16 Kirk, Raven and Schofield, *The Presocratic Philosophers*, p. 154.
17 Karl Popper, *Conjectures and Refutations: The Growth of Scientific Knowledge* (London, 2002), p. 185.
18 Ibid., p. 186.
19 Phillip Sidney Horky, 'When Did *Kosmos* Become the *Kosmos?*', in *Cosmos in the Ancient World*, ed. Phillip Sidney Horky (Cambridge, 2019), p. 23.
20 He's also supposed to have measured the period between the longest day (called the summer solstice) and shortest (the winter solstice), as well as the spring and autumn equinoxes – when day and night are the same length. See Graham, *Science before Socrates*, p. 50.
21 D. R. Dicks, *Early Greek Astronomy to Aristotle* (London, 1970), p. 32.
22 Pseudo-Plato, 'Epinomis', in *Plato: Charmides et al.*, trans. W.R.M. Lamb (Cambridge MA, 1927), p. 471 (987b).
23 Dicks, *Early Greek Astronomy to Aristotle*, pp. 165–7.
24 Things were not so simple in the ancient world. The Earth's axis wobbles: a phenomenon called the 'procession of the equinoxes' discovered by a Greek astronomer called Hipparchus in around 150 BC. This means that the pole star isn't fixed in the place it is today, but instead ambles around the pole. The ancient Greeks had to use the entire constellation 'the Little Bear', in which the pole star sits, as a general indication of which direction north was in.

6 The Presocratics and Socrates: 'Floating on air'

1 Daniel W. Graham, *Science before Socrates: Parmenides, Anaxagoras, and the New Astronomy* (Oxford, 2013), p. 96.
2 D. R. Dicks, *Early Greek Astronomy to Aristotle* (London, 1970), p. 51.
3 G. S. Kirk, J. E. Raven and M. Schofield, *The Presocratic Philosophers: A Critical History with a Selection of Texts* (Cambridge, 1983), p. 252.
4 Dirk Couprie, 'Some Remarks on the Earth in Plato's *Phaedo*', *Hyperboreus*, XI (2005), p. 194.
5 A surviving account of Parmenides' thought places the Earth at the centre of concentric rings of light and darkness, which is a vision in two dimensions rather than three. This seems to militate against the Earth being a sphere. See Kirk, Raven and Schofield, *The Presocratic Philosophers*, p. 258.

6 Graham, *Science before Socrates*, p. 91.
7 One source attributes this idea to the Milesian Anaximenes, but it conflicts with the rest of his doctrine as we understand it. See Graham, *Science before Socrates*, p. 65.
8 Kirk, Raven and Schofield, *The Presocratic Philosophers*, p. 259.
9 Plato, 'Parmenides', in *Plato: Cratylus and Others*, trans. Harold North Fowler (Cambridge, MA, 1926), p. 201 (127a).
10 Kirk, Raven and Schofield, *The Presocratic Philosophers*, p. 381.
11 Diogenes Laertius, *Lives of Eminent Philosophers*, ed. R. D. Hicks (Cambridge, MA, 1925), vol. I, p. 141 (2.3.10).
12 Plato, 'Cratylus', in *Plato: Cratylus and Others*, trans. Harold North Fowler (Cambridge MA, 1926), p. 91 (409a).
13 Graham, *Science before Socrates*, p. 124.
14 Cicero, 'On the Republic', in *Cicero: On the Republic, On the Laws*, trans. Clinton W. Keyes (Cambridge, MA, 1928), p. 45 (1:16).
15 Plutarch, 'Life of Pericles', in *The Rise and Fall of Athens: Nine Greek Lives*, ed. Ian Scott-Kilvert (Harmonsworth, 1960), p. 201 (35).
16 James Hannam, *God's Philosophers: How the Medieval World Laid the Foundations of Modern Science* (London, 2009), p. 340.
17 Aristotle, 'Meteorology', in *The Complete Works of Aristotle*, ed. Jonathan Barnes (Princeton, NJ, 1984), p. 591 (365a).
18 Kirk, Raven and Schofield, *The Presocratic Philosophers*, p. 385.
19 Ibid., p. 387. Presumably, Archelaus believed that the edge of the bowl cast a shadow as the Sun rose above its eastern rim. However, that would mean its light became visible to people on the western side of the world before those in the east, which is the opposite of what we observe.
20 Kirk, Raven and Schofield, *The Presocratic Philosophers*, p. 387.
21 Aristophanes, 'The Clouds', in *Lysistrata and Other Plays*, trans. Alan Sommerstein (Harmondsworth, 1973), p. 128 (390).
22 Ibid., p. 116 (94).
23 Ibid., p. 123 (260).
24 Anthony Kenny, *A New History of Western Philosophy* (Oxford, 2010), p. 34.

7 Plato: 'Flat or round, whichever is better'

1 Alfred North Whitehead, *Process and Reality: An Essay in Cosmology* (New York, 1978), p. 39.
2 Anthony Kenny, *A New History of Western Philosophy* (Oxford, 2010), p. 43.
3 G. S. Kirk, J. E. Raven and M. Schofield, *The Presocratic Philosophers: A Critical History with a Selection of Texts* (Cambridge, 1983), pp. 214–15.
4 Walter Burkert, *Lore and Science in Ancient Pythagoreanism* (Cambridge, MA, 1972), p. 217.

5 Kirk, Raven and Schofield, *The Presocratic Philosophers*, p. 230.
6 Aristotle, 'On the Heavens', in *The Complete Works of Aristotle*, ed. Jonathan Barnes (Princeton, NJ, 1984), p. 479 (291a).
7 Laertius, *Lives of Eminent Philosophers*, trans. R. D. Hicks (Cambridge, MA, 1925), vol. II, p. 343 (8.1.26) and p. 365 (8.1.49).
8 Ibid., vol. I, p. 131 (2.1).
9 Diogenes did provide references to books he'd read that are now lost, but these are known to date from centuries after the death of Pythagoras. One of the few modern scholars to take Diogenes Laertius at face value was Sir Thomas Heath (1861–1940), a distinguished historian of Greek science, who repeated the statement that the Globe began with Pythagoras. See T. L. Heath, *Aristarchus of Samos: The Ancient Copernicus* (Cambridge, 1913), p. 51. Heath's influence is such that his claim remains in circulation today, widely repeated in books and on the Internet. In contrast, from the second half of the twentieth century, experts on ancient philosophy have become increasingly sceptical that we can be sure of anything about Pythagoras. See D. R. Dicks, *Early Greek Astronomy to Aristotle* (London, 1970), p. 64.
10 The fragments of Philolaus, with a copious commentary, are set out in Carl A. Huffman, *Philolaus of Croton: Pythagorean and Presocratic. A Commentary on the Fragments and Testimonia with Interpretive Essays* (Cambridge, 1993).
11 Kirk, Raven and Schofield, *The Presocratic Philosophers*, p. 340.
12 Aristotle, 'On the Heavens', p. 482 (293a).
13 Ibid., p. 483 (293b).
14 George Bosworth Burch, 'The Counter-Earth', *Osiris*, XI (1954), p. 273.
15 Aristotle, *Metaphysics*, trans. Hugh Lawson-Tancred (Harmondsworth, 1998), p. 20 (986a).
16 Aristotle, 'On the Heavens', p. 483 (291b).
17 Huffman, *Philolaus of Croton*, p. 5.
18 There is a contested fragment (F17) attributed to Philolaus that asserts that upwards and downwards are equivalent and bear the same relation to the centre. If genuine, the fragment would be evidence that he had grasped that the Globe requires a redefinition of the meaning of 'down'. See Huffman, *Philolaus of Croton*, p. 215.
19 Plato, 'Phaedo', in *The Last Days of Socrates*, trans. Hugh Tredennick and Harold Tarrant (Harmondsworth, 1993), p. 111 (59b).
20 Xenophon, 'Memoires of Socrates', in *Conversations of Socrates*, trans. Robin Waterfield and Hugh Tredennick (Harmondsworth, 1990), p. 71 (1.1.14).
21 Plato, 'Phaedo', p. 161 (97d).
22 Daniel W. Graham, *Science before Socrates: Parmenides, Anaxagoras, and the New Astronomy* (Oxford, 2013), p. 96. In the fifth century AD

Martianus Capella attributed an argument to Anaxagoras saying
the Sun and Moon come immediately into view when they rise
above the horizon, which he took as proof that the Earth is flat.
Panchenko supposes Anaxagoras is arguing against the Globe, so
must have been aware of it (see Dimitri Panchenko, 'Anaxagoras'
Argument against the Sphericity of the Earth', *Hyperboreus*, III/1
(1997), p. 177). However, this does not appear to be the case. The
argument echoes the one attributed to Archelaus by Hippolytus,
which doesn't have anything to do with the Globe (see Kirk, Raven
and Schofield, *The Presocratic Philosophers*, p. 387). Instead, the
dispute referenced by Martianus Capella seems to be over whether
the Earth is flat (as Anaxagoras thought) or shaped like a bowl (as
Archelaus said).
23 Huffman, *Philolaus of Croton*, p. 5.
24 Plato, 'Timaeus', in *Timaeus and Critias*, trans. Desmond Lee
(Harmondsworth, 1977), p. 42 (29c).
25 Plato presented alternative and equally fabulous cosmologies in
some of his other works. The most famous is the myth of Ur at the
end of the Republic. Plato concluded this dialogue with the tale of
a man who visits the realm of the dead, sees the universe from the
outside, and then returns to his body to tell his companions what
he saw. It makes no reference to the shape of the Earth.
26 Plato, 'Phaedo', p. 175 (108e).
27 Dirk Couprie, 'Some Remarks on the Earth in Plato's *Phaedo*',
Hyperboreus, XI (2005), p. 198.
28 Plato, 'Phaedo', p. 181 (114d).
29 Plato, 'Phaedrus', in *Phaedrus and Letters VII and VIII*, trans. Walter
Hamilton (Harmondsworth, 1973), p. 52 (247b).

8 Aristotle: 'Necessarily spherical'

1 Plato, *Theaetetus*, ed. Robin A. H. Waterfield (Harmondsworth,
1987), p. 115 (201c).
2 I have resisted the temptation to see if my account of the discovery
of the Globe can be shoehorned into any of the philosophical
models put forward by the likes of Karl Popper and Thomas
Kuhn. While falsification and paradigm shifts can be useful to
historians as lenses to examine scientific developments, reality
is too complicated to fit one-dimensional theories. That said, my
personal loyalty should be obvious enough to those who care about
such things.
3 Diogenes Laertius, *Lives of Eminent Philosophers*, ed. R. D. Hicks
(Cambridge, MA, 1925), vol. II, p. 405.
4 George Huxley, 'Studies in the Greek Astronomers', *Greek, Roman,
and Byzantine Studies*, IV/2 (1963), pp. 83–7.
5 Eratosthenes and Hyginus, *Constellation Myths with Aratus's
Phaenomena*, ed. Robin Hard (Oxford, 2015), pp. 137–67.

6 Ibid., p. 154.

7 Huxley, 'Studies in the Greek Astronomers', p. 88.

8 Plato, *The Laws*, trans. Trevor J. Saunders (Harmondsworth, 1975), p. 316 (821a).

9 D. R. Dicks, *Early Greek Astronomy to Aristotle* (London, 1970), p. 108.

10 Aristotle, *Metaphysics*, trans. Hugh Lawson-Tancred (Harmondsworth, 1998), p. 378 (1074b).

11 Ibid., p. 176.

12 Plato, *The Laws*, p. 317 (822a).

13 Quintus Curtius Rufus, *The History of Alexander*, ed. John Yardley and Waldemar Heckel (Harmondsworth, 1984), p. 217 (9.2.26).

14 Arrian, *The Campaigns of Alexander*, ed. Aubrey de Sélincourt and J. R. Hamilton (Harmondsworth, 1971), p. 293 (5.26).

15 Carlo Natali, *Aristotle* (Princeton, NJ, 2013), p. 62.

16 Cicero, 'Topica', in *Cicero: On Invention. The Best Kind of Orator. Topics*, trans. H. M. Hubbell (Cambridge, MA, 1949), p. 385 (3).

17 As a matter of fact, it is possible to see direct evidence that the Earth is rotating. A weight hung from a long pendulum remains unaffected while the Earth turns beneath it. As the pendulum swings backwards and forwards in a plane, it appears that the pendulum's swing is swivelling. In reality, it is the Earth that is rotating as the pendulum swings in place. Frenchman Léon Foucault first erected such a pendulum in 1851. You can still see a replica of it in the Panthéon in Paris, and there are now many others in museums around the world. See Harold L. Burstyn, 'Foucault, Jean Barnard Léon', in *Dictionary of Scientific Biography*, ed. Charles Coulston Gillispie (New York, 1970), vol. v, p. 86.

18 Aristotle, 'On the Heavens', in *The Complete Works of Aristotle*, ed. Jonathan Barnes (Princeton, NJ, 1984), p. 486 (295b).

19 Ibid., p. 487 (296b).

20 Aristotle, *Politics*, ed. T. A. Sinclair and Trevor J. Saunders (Harmondsworth, 1981), p. 69 (1254b).

21 The author thanks Peter Gainsford for this insight.

22 Aristotle, 'On the Heavens', p. 450 (270b).

23 John Losee, *A Historical Introduction to the Philosophy of Science* (Oxford, 2001), p. 5.

24 Jonathan Barnes, *Aristotle* (Oxford, 1982), p. 32.

25 Aristotle, 'On the Heavens', p. 488 (297a).

26 Ibid., p. 489 (297b).

27 Ibid., p. 489 (298a).

28 J. M. Bigwood, 'Aristotle and the Elephant Again', *American Journal of Philology*, CXIV/4 (1993), p. 547.

29 Eudoxus wrote a geography book called *A Circuit of the Earth*. See James S. Romm, *The Edges of the Earth in Ancient Thought: Geography, Exploration, and Fiction* (Princeton, NJ, 1992), p. 26.

30 E. L. Gettier, 'Is Justified True Belief Knowledge?', *Analysis*, XXIII (1963), pp. 121–3.
31 James Hannam, *God's Philosophers: How the Medieval World Laid the Foundations of Modern Science* (London, 2009), p. 171.

9 Greek Debate on the Shape of the World: 'Either round or triangular or some other shape'

1 Roger S. Bagnall, 'Alexandria: Library of Dreams', *Proceedings of the American Philological Society*, CXLVI/4 (2002), pp. 348–62.
2 There is a great deal of misinformation about Eratosthenes' estimate. For a breakdown of some misconceptions, see Peter Gainsford, 'The Eratosthenes Video Published by Business Insider: A Fact-Check', http://kiwihellenist.blogspot.com, 2016. Note that Eratosthenes was able to glean all the information he needed from government records in Alexandria, so was very much an armchair scholar. We should not imagine he trekked all the way to Syene and Meroë to measure any angles for himself.
3 Cleomedes, *Cleomedes' Lectures on Astronomy: A Translation of the Heavens, ed.* Alan C. Bowen and Robert B. Todd (Berkeley, CA, 2004), p. 82 (1.7).
4 If the Earth was actually flat, simple trigonometry allows us to calculate that his observations would have implied that the Sun was less than 7,240 kilometres (4,500 mi.) away.
5 Ptolemy of Alexandria, *Ptolemy's Almagest*, ed. G. J. Toomer (London, 1984), p. 41 (1.4).
6 Diogenes Laertius, *Lives of Eminent Philosophers*, trans. R. D. Hicks (Cambridge, MA, 1925), vol. II, p. 111 (7.1.2).
7 Samuel Sambursky, *Physics of the Stoics* (Princeton, NJ, 1959), p. 108.
8 Cleomedes, *Cleomedes' Lectures on Astronomy and Geminos, Geminos's Introduction to the Phenomena: A Translation and Study of a Hellenistic Survey of Astronomy*, ed. James Evans and J. L. Berggren (Princeton, NJ, 2006).
9 David Furley, 'Cosmology', in *The Cambridge History of Hellenistic Philosophy*, ed. Keimpe Algra et al. (Cambridge, 2005), p. 421. More recently, Frederik Bakker made a heroic effort to show the Epicureans did not believe the Earth was flat, but his detailed arguments depend on finding reasons to deny the plain meaning of the surviving texts. See Frederik A. Bakker, *Epicurean Meteorology: Sources, Method, Scope and Organization* (Leiden, 2016), pp. 162–263.
10 Aristotle, 'On the Heavens', in *The Complete Works of Aristotle*, ed. Jonathan Barnes (Princeton, NJ, 1984), p. 484 (294b).
11 Aristotle, 'On Generation and Corruption', in *The Complete Works of Aristotle*, p. 533 (326a).
12 Diogenes Laertius, *Lives of Eminent Philosophers*, vol. II, p. 537 (10.9).

13 Pliny the Younger, *The Letters of the Younger Pliny*, trans. Betty Radice (Harmondsworth, 1969), p. 171 (6.20).

14 I. Bukreeva et al., 'Virtual Unrolling and Deciphering of Herculaneum Papyri by X-ray Phase-Contrast Tomography', *Scientific Reports*, VI (2016).

15 *The Epicurus Reader: Selected Writings and Testimonia*, ed. and trans. Brad Inwood and L. P. Gerson (Indianapolis, IN, 1994), p. 29 (10.125).

16 Ibid., p. 20 (10.88).

17 Ibid., p. 21 (10.91).

18 Carlo Natali, *Aristotle* (Princeton, NJ, 2013), p. 9.

19 Lucretius, *On the Nature of the Universe*, trans. R. E. Latham and John Godwin (London, 2005), p. 251.

20 Cicero, *Letters to Friends*, trans. D. R. Shackleton Bailey (Cambridge, MA, 2001), vol. I, p. 267 (63/13.1).

21 D. N. Sedley, *Lucretius and the Transformation of Greek Wisdom* (Cambridge, 1998), p. 92.

22 Lucretius, *On the Nature of the Universe*, p. 145 (5.638).

23 Ibid., p. 36 (1.1061); Furley, 'Cosmology', p. 421.

24 James Warren, 'Lucretius and Greek Philosophy', in *The Cambridge Companion to Lucretius*, ed. Stuart Gillespie and Philip R. Hardie (Cambridge, 2007), p. 23.

25 Cicero, *The Nature of the Gods*, trans. H.C.P. McGregor (Harmondsworth, 1972), p. 142 (2.48).

26 See, for example, Matt Ridley, *The Evolution of Everything* (London, 2016).

27 *The Epicurus Reader*, p. 17 (10:80).

28 Julian, 'Fragment of a Letter to a Priest', in *Julian*, trans. Wilmer C. Wright (Cambridge, MA, 1913), vol. II, p. 327 (301c).

29 Suetonius, 'On Grammarians', in *Suetonius*, trans. J. C. Rolfe (Cambridge, MA, 1914), vol. II, p. 383 (2).

30 Geminos, *Geminos's Introduction to the Phenomena*, p. 215 (16.28).

31 James S. Romm, *The Edges of the Earth in Ancient Thought: Geography, Exploration, and Fiction* (Princeton, NJ, 1992), p. 180.

32 Ibid., p. 188.

33 G. S. Kirk, J. E. Raven and M. Schofield, *The Presocratic Philosophers: A Critical History with a Selection of Texts* (Cambridge, 1983), p. 104.

34 Herodotus, *The Histories*, trans. Aubrey de Sélincourt (Harmondsworth, 1996), p. 227 (4.36).

35 Ibid., p. 228 (4.42).

36 Aristotle, 'Meteorology', in *The Complete Works of Aristotle*, p. 587 (362b).

37 The modern definition of the Arctic and Antarctic Circles is the latitude furthest from the pole at which the Sun is still visible at any time on the day of the winter solstice.

38 Aristotle, 'Meteorology', p. 587 (362b).
39 Geminus, *Geminos's Introduction to the Phenomena*, p. 215 (16.20).
40 Ptolemy of Alexandria, *Ptolemy's Geography: An Annotated Translation of the Theoretical Chapters*, trans. J. L. Berggren and Alexander Jones (Princeton, NJ, 2000), p. 21.

10 Romans on the Globe: 'The circle of the world'

1   As Roman emperors shed the pretence that they were first among equals in the Senate and asserted their supreme power, they were more willing to portray themselves in cosmic terms. So, it's not always clear whether an orb on a coin represents the Earth or the celestial sphere. Where the ecliptic (the path of the Sun across the stars) is marked on the sphere, we can be sure it symbolizes the heavens.
2   Raymond V. Sidrys, *The Mysterious Spheres on Greek and Roman Ancient Coins* (Oxford, 2020), p. 96.
3   Databases of papyrus fragments such as *Trismegistos* are in the public domain, so we can see for ourselves what has been discovered. See www.trismegistos.org, 8 July 2022.
4   Alexander Jones, *Astronomical Papyri from Oxyrhynchus* (Philadelphia, PA, 1999), p. 4.
5   Reviel Netz, 'The Bibliosphere of Ancient Science (Outside of Alexandria): A Preliminary Survey', *Naturwissenschaften, Technik und Medizin*, XIX/3 (2011), p. 248.
6   Horace, 'Epistles', in *Satires. Epistles. The Art of Poetry*, trans. H. Rushton Fairclough (Cambridge, MA, 1926), p. 409 (2.1.156).
7   Carl Sagan, *Pale Blue Dot* (London, 1994).
8   Cicero, 'The Republic', in *Cicero: On the Republic, on the Laws*, trans. Clinton W. Keyes (Cambridge, MA, 1928), p. 275 (6.22).
9   Cicero, *The Nature of the Gods*, trans. H.C.P. McGregor (Harmondsworth, 1972), p. 142 (2.47). For the Latin, see Cicero, 'De natura deorum', in *On the Nature of the Gods. Academics*, ed. H. Rackham (Cambridge, MA, 1933), p. 168.
10 To make matters worse, the Latin word *mundus*, usually translated as 'world', can refer either to the Earth or to the entire universe. So, referring to the *mundus* being spherical could refer to the universe, rather than Earth, being ball-shaped.
11 Pliny the Younger, *The Letters of the Younger Pliny*, ed. Betty Radice (Harmondsworth, 1969), p. 169 (6.16).
12 Pliny the Elder, *Natural History*, trans. H. Rackham (Cambridge, MA, 1938), vol. I, p. 295 (2.64).
13 Ibid., p. 299 (2.65).
14 Ibid., p. 311 (2.71); ibid., p. 319 (2.77).
15 Ibid., p. 315 (2.73).
16 Ibid., p. 297 (2.65). Varro, a prolific Roman author of the first century BC, had said it was 'egg-shaped' rather than a sphere.

He's quoted in Cassiodorus, *Institutions of Divine and Secular Learning*, trans. James W. Halporn and Mark Vessey (Liverpool, 2004), p. 229 (2.7.4). Unfortunately, very little of his work is extant, so it is not clear what he meant by this except that the surface of the Earth is curved.

17 Virgil, 'Georgics', in *Eclogues. Georgics. Aeneid: Books 1–6*, trans. H. Rushton Fairclough (Cambridge, MA, 1916), p. 97 (1.231).

18 Both ancient and modern commentators have accused Virgil of being sloppy about astronomy on the strength of this passage. See Frederik A. Bakker, 'Vergilius Astronomiae Ignarus? A Vindication of Virgil's Astronomical Knowledge in *Georgics* 1.231–258', *Mnemosyne*, LXXII/4 (2019), pp. 621–46.

19 Ovid, *Metamorphoses*, trans. Mary Innes (Harmondsworth, 1955), p. 30 (1.35).

20 Lucan, *The Civil War*, ed. J. D. Duff (Cambridge, MA, 1928), vol. III, p. 571 (9.878).

21 Macrobius, *Commentary on the Dream of Scipio*, trans. William Harris Stahl (New York, 1952), pp. 154 and 172 (1.16.10 and 1.20.20). The difference between the two commonly cited figures for Eratosthenes' estimate of 250,000 stades and 252,000 stades may result from his calculations based on the Sun being at an infinite distance from the Earth and 4,080,000 stades away respectively. See Christián Carlos Carman and James Evans, 'The Two Earths of Eratosthenes', *Isis*, CVI/1 (2015), pp. 1–16. Alternatively, 252,000 might be preferred because it is divisible by 60.

22 Martianus Capella, *The Marriage of Philology and Mercury*, trans. William Harris Stahl and Richard Johnson (New York, 1977), p. 220 (6.590).

11 India: 'The mountain at the North Pole'

1 K. V. Sarma, 'Lalla', in *Encyclopaedia of the History of Science, Technology, and Medicine in Non-Western Cultures*, ed. Helaine Selin (Dordrecht, 1997), p. 508.

2 Bidare V. Subbarayappa and K. V. Sarma, ed., *Indian Astronomy: A Source-Book* (Bombay, 1985), pp. 41 and 44.

3 The Sanskrit word for tortoise and turtle is the same, so the species of the reptile Lalla's opponents said held up the world is ambiguous.

4 See chapters Eleven and Twelve of R. S. Sharma, *India's Ancient Past* (New Delhi, 2005) and Chapter Four of Romila Thapar, *Early India: From the Origins to AD 1300* (Berkeley, CA, 2002).

5 Wendy Doniger O'Flaherty, trans., *The Rig Veda* (Harmondsworth, 1981), p. 211 (5.85.1).

6 Ibid., p. 203 (1.160).

7 Juan Mascaró, trans., *The Bhagavad Gita* (Harmondsworth, 1962), p. 46 (1.35).

8 *The Rig Veda*, p. 28 (10.121.5).
9 Ibid., p. 204 (1.185).
10 Richard F. Gombrich, 'Ancient Indian Cosmology', in *Ancient Cosmologies*, ed. Carmen Blacker and Michael Loewe (London, 1975), p. 118.
11 Kim Plofker, 'Astronomy and Astrology on India', in *The Cambridge History of Science*, vol. I: *Ancient Science*, ed. Alexander Jones and Liba Taub (Cambridge, 2003), p. 486.
12 *The Rig Veda*, p. 188 (5.40.5).
13 Subbarayappa and Sarma, *Indian Astronomy: A Source-Book*, p. 1.
14 Kim Plofker, *Mathematics in India* (Princeton, NJ, 2009), p. 41.
15 See chapters Sixteen and Eighteen of Sharma, *India's Ancient Past* and chapters Five and Six of Thapar, *Early India*.
16 Plofker, *Mathematics in India*, p. 52; David Pingree, 'The Purāṇas and Jyotiḥśāstra: Astronomy', *Journal of the American Oriental Society*, CX/2 (1990), p. 275.
17 *Hindu Myths*, trans. Wendy O'Flaherty (Harmondsworth, 1975), pp. 274–80 (1.16). The Sanskrit word 'Ketu' could also mean a light in the sky such as a comet or shooting star.
18 Devabrata M. Bose, Samarendra Nath Sen and Bidare V. Subbarayappa, *A Concise History of Science in India* (New Delhi, 1971), p. 65.
19 Plofker, *Mathematics in India*, p. 67.
20 Adam Bowles, trans., *The Mahabharata* VIII (New York, 2006), p. 447 (8.45.35).
21 Bose, Sen and Subbarayappa, *A Concise History of Science in India*, p. 79.
22 R. C. Gupta, 'Aryabhata', in *Encyclopaedia of the History of Science, Technology, and Medicine in Non-Western Cultures*, ed. Helaine Selin (Dordrecht, 1997), p. 72.
23 Aryabhata, *The Aryabhatiya of Aryabhata*, trans. David Eugene Smith (Chicago, IL, 1930), p. 64 (4.6).
24 Ibid., p. 68 (4.11).
25 Ibid., pp. 64–6 (4.9–10).
26 Subbarayappa and Sarma, *Indian Astronomy: A Source-Book*, p. 44.
27 Plofker, *Mathematics in India*, p. 114.
28 Ibid., p. 115.
29 Ibid., p. 50.
30 David Pingree, *The Yavanajātaka of Sphujidhvaja* (Cambridge, MA, 1978).
31 Strabo, *Geography*, trans. Horace Leonard Jones (Cambridge, MA, 1930), vol. VII, p. 103 (15.1.59).
32 Barry Cunliffe, *By Steppe, Desert, and Ocean: The Birth of Eurasia* (Oxford, 2015), pp. 290–92.
33 Ibid., pp. 264–5.

34 Greek shipping in the Indian Ocean did not hug the coast but sailed directly across to the Malabar Coast before the monsoon winds. See *The Periplus of the Erythraean Sea*, ed. G.W.B. Huntingford (London, 1980), p. 52 (57).

35 Plofker, *Mathematics in India*, p. 48.

36 Pingree, 'The Purāṇas and Jyotiḥśāstra: Astronomy', p. 279.

37 Ibid., p. 276.

38 Christopher Minkowski, 'Competing Cosmologies in Early Modern Indian Astronomy', in *Ketuprakāśa: Studies in the History of the Exact Sciences in Honor of David Pingree*, ed. Charles Burnett, Jan Hogendijk and Kim Plofker (Leiden, 2004), p. 360.

39 Toke Lindegaard Knudsen, *The Siddhāntasundara of Jñānarāja, an English Translation with Commentary* (Baltimore, MD, 2014), pp. 51 and 54.

40 Ibid., pp. 55–7.

41 Minkowski, 'Competing Cosmologies in Early Modern Indian Astronomy', p. 381.

42 Kim Plofker, 'Derivation and Revelation: The Legitimacy of Mathematical Models in Indian Cosmology', in *Mathematics and the Divine: A Historical Study*, ed. T. Koetsier and L. Bergmans (Amsterdam, 2004), p. 72.

12 The Sassanian Persians: 'Good thoughts, good words, good deeds'

1 Touraj Daryaee, 'Mind, Body, and the Cosmos: Chess and Backgammon in Ancient Persia', *Iranian Studies*, XXXV/4 (2002), pp. 281–312.

2 Touraj Daryaee, *Sasanian Persia: The Rise and Fall of an Empire* (London, 2009), p. 81.

3 Ibid., p. 86.

4 Zsuzsanna Gulácsi and Jason BeDuhn, 'Picturing Mani's Cosmology: An Analysis of Doctrinal Iconography on a Manichaean Hanging Scroll from 13th/14th-Century Southern China', *Bulletin of the Asia Institute*, 25 (2011), pp. 55–105.

5 Alan Cameron, 'The Last Days of the Academy at Athens', *Proceedings of the Cambridge Philological Society*, 15 (1969), p. 8.

6 Kevin van Bladel, 'The Arabic History of Science of Abū Sahl Ibn Nawbaḫt (*fl.* ca 770–809) and Its Middle Persian Sources', in *Islamic Philosophy, Science, Culture, and Religion*, ed. Felicitas Opwis and David Reisman (Leiden, 2012), p. 46.

7 Daryaee, *Sasanian Persia*, p. 120.

8 Van Bladel, 'The Arabic History of Science of Abū Sahl', p. 47.

9 David Frendo, 'Agathias' View of the Intellectual Attainments of Khusrau I: A Reconsideration of the Evidence', *Bulletin of the Asia Institute*, 18 (2004), pp. 97–100.

10 Priscian, *Answers to King Khosroes of Persia*, trans. Pamela Huby (London, 2016), p. 42.

11 Daryaee, *Sasanian Persia*, p. 119.

12 While there are no extant citations from a Persian version of Ptolemy's *Almagest*, the 'Royal Almanac' commissioned by the shahs did utilize Indian and Ptolemaic parameters. See Emily Cottrell and Micah Ross, 'Persian Astrology: Dorotheus and Zoroaster According to the Medieval Arabic Sources (8th–11th Century.)', in *Proceedings of the 8th European Conference of Iranian Studies* (St Petersburg, 2019), p. 90.

13 Frendo, 'Agathias' View of the Intellectual Attainments of Khusrau I', p. 99.

13 Early Judaism: 'From the ends of the Earth'

1 L. Miller and Maurice Simon, trans., 'Bekoroth', in *The Babylonian Talmud*, ed. I. Epstein (London, 1948), pp. 51–4 (8b).

2 Martin Goodman, *A History of Judaism* (London, 2019), p. 263.

3 John H. Walton, *Ancient Near Eastern Thought and the Old Testament: Introducing the Conceptual World of the Hebrew Bible* (Grand Rapids, MI, 2018), p. 133.

4 Daniel Harlow, 'Creation According to Genesis: Literary Genre, Cultural Context, Theological Truth', *Christian Scholar's Review*, XXXVII/2 (2008), pp. 163–98.

5 M. A. Knibb, trans., '1 Enoch', in *The Apocryphal Old Testament*, ed. H.F.D. Sparks (Oxford, 1984), p. 257 (1 Enoch 72).

6 J. Edward Wright, *The Early History of Heaven* (New York, 2000), p. 129.

7 Ibid., pp. 137 and 154.

8 Moshe Simon-Shoshan, 'The Heavens Proclaim the Glory of God: A Study in Rabbinic Cosmology', *Bekhol Derakhekha Daehu – Journal of Torah and Scholarship*, 20 (2008), p. 73.

9 Ibid., p. 78.

10 Ibid., p. 91.

11 H. Freedman trans., 'Pesahim', in *The Babylonian Talmud*, ed. I. Epstein (London, 1938), p. 505 (94b:5).

12 H. St J. Thackeray, *The Letter to Aristeas: Translated with an Appendix of Ancient Evidence of the Origin of the Septuagint* (London, 1917), p. 9.

13 Philo, 'A Treatise on the Cherubim', in *Philo*, ed. F. H. Colson and G. H. Whitaker (Cambridge, MA, 1929), vol. II, p. 21 (7).

14 Philo, *Questions of Genesis*, ed. Ralph Marcus (Cambridge, MA, 1953), p. 52 (1.84).

15 Josephus, *Jewish Antiquities*, trans. H. St J. Thackeray (Cambridge, MA, 1930), pp. 15 and 19 (1.31 and 1.38).

16 Goodman, *A History of Judaism*, p. 281.

17 It was Christians who preserved the writings of Philo, finding that they shed light on obscure passages in the Bible. It was not until the

sixteenth century that Philo was reintroduced to Jewish readers. See Goodman, *A History of Judaism*, p. 366.

14 Christianity: 'All things established by divine command'

1 John P. Meier, *A Marginal Jew: The Roots of the Problem and the Person* (New York, 1991), p. 402.

2 Larry Siedentop, *Inventing the Individual* (London, 2015), p. 14.

3 For example, the Shorter Testament of Abraham and the Apocalypse of Paul. See J. Edward Wright, *The Early History of Heaven* (New York, 2000), pp. 154 and 162.

4 David Lindberg, 'Science and the Early Church', in *God and Nature: Historical Essays on the Encounter between Science and Christianity*, ed. David Lindberg and Ronald Numbers (Berkeley, CA, 1986).

5 Basil of Caesarea, 'Hexaemeron', in *St Basil: Letters and Select Works*, ed. Philip Schaff and Henry Wace (New York, 1895), p. 57 (1.10).

6 Ibid., p. 83 (6.3).

7 John of Damascus, 'Exposition of the Orthodox Faith', in *St Hilary of Poitiers and John of Damascus*, ed. S.D.F. Salmond (Oxford, 1899), p. 22 (2.6).

8 Ibid., p. 25 (2.7).

9 Ibid., p. 29 (2.10).

10 Photius, *The Bibliotheca*, ed. Nigel Wilson (London, 1994), p. 214.

11 Hervé Inglebert, '"Inner" and "Outer" Knowledge: The Debate between Faith and Reason in Late Antiquity', in *A Companion to Byzantine Science*, ed. Stavros Lazaris (Leiden, 2020), pp. 27–52 (p. 46).

12 John Chrysostom, 'Homilies of the Epistle to the Hebrews', in *Homilies on the Gospel of St John and the Epistle to the Hebrews*, ed. Philip Schaff (New York, 1889), p. 433 (14.1).

13 Philip Jenkins, *The Lost History of Christianity* (New York, 2008), p. 58.

14 Said Hayati, 'Mar Aba I: Historical Context and Biographical Reconstruction', MA dissertation, University of Salzburg, 2018, p. 27.

15 Birgitta Elweskiöld, 'John Philoponus against Cosmas Indicopleustes: A Christian Controversy on the Structure of the World in Sixth-Century Alexandria', PhD thesis, Lund University, 2005, p. 15.

16 J. W. McCrindle, ed., *The Christian Topography of Cosmas* (London, 1897), p. 25.

17 Travis Lee Clark, 'Imaging the Cosmos: The Christian Topography by Kosmas Indikopleustes', PhD thesis, Temple University, 2008, p. 10.

18 Elweskiöld, 'John Philoponus against Cosmas Indicopleustes', p. 8.

19 McCrindle, *The Christian Topography*, p. 129.
20 Ibid., p. 17.
21 Ibid., p. 347.
22 Elweskiöld, 'John Philoponus against Cosmas Indicopleustes', p. 109.
23 Richard Sorabji, 'John Philoponus', in *Philoponus and the Rejection of Aristotelian Science*, ed. Richard Sorabji (London, 2010), p. 47.
24 Maja Kominko, *The World of Kosmas* (Cambridge, 2013), p. 19.
25 Elweskiöld, 'John Philoponus against Cosmas Indicopleustes', p. 94.
26 Kevin van Bladel, 'Heavenly Cords and Prophetic Authority in the Quran and Its Late Antique Context', *Bulletin of the School of Oriental and African Studies*, LXX/2 (2007), p. 226.
27 Robert Hewson, 'Science in Seventh-Century Armenia: Ananias of Sirak', *Isis*, LIX/1 (1968), p. 41.
28 Severus Sebokht, 'Description of the Astrolabe', in *Astrolabes of the World*, ed. R. T. Gunther (Oxford, 1932).
29 Photius, *The Bibliotheca*, p. 31. Obviously, Photius did not realize Cosmas had based his work on a world picture held by John Chrysostom, his esteemed predecessor as Archbishop of Constantinople.
30 Jeffrey Burton Russell, *Inventing the Flat Earth: Columbus and Modern Historians* (Westport, VA, 1991), p. 34.
31 Anne-Laurence Caudano, 'Un Univers sphérique ou voûté? Survivance de la cosmologie Antiochienne à Byzance (XIe et XIIe S.)', *Byzantion*, LXXVIII (2008), p. 71.
32 Cyril A. Mango, *Byzantium: The Empire of New Rome* (London, 1980), p. 176.
33 Anne Tihon, 'Astronomy', in *The Cambridge Intellectual History of Byzantium*, ed. Anthony Kaldellis and Niketas Siniossoglou (Cambridge, 2017), p. 186.

15 Islam: 'The Earth laid out like a carpet'

1 David Pingree, 'The Fragments of the Works of Al-Fazārī', *Journal of Near Eastern Studies*, XXIX/2 (1970), p. 104.
2 S. Frederick Starr, *Lost Enlightenment: Central Asia's Golden Age from the Arab Conquest to Tamerlane* (Princeton, NJ, 2013), pp. 119–24.
3 Dimitri Gutas, *Greek Thought, Arabic Culture: The Graeco-Arabic Translation Movement in Baghdad and Early Abbāsid Society (2nd–4th/8th–10th Centuries)* (London, 1998), p. 67.
4 Kevin van Bladel, 'Eighth-Century Indian Astronomy in the Two Cities of Peace', in *Islamic Cultures, Islamic Contexts: Essays in Honor of Professor Patricia Crone*, ed. Behnam Sadeghi et al. (Leiden, 2014), p. 266.
5 Fitzroy Morrissey, *A Short History of Islamic Thought* (London, 2021), p. 54.

6 Ṣāʿid ibn Aḥmad Andalusī, *Science in the Medieval World: Book of the Categories of Nations*, trans. Semaʿan I. Salem and Alok Kumar (Austin, TX, 1991), p. 46.

7 Seb Falk, *The Light Ages: A Medieval Journey of Discovery* (London, 2020), p. 242.

8 Pingree, 'The Fragments of the Works of Al-Fazārī', p. 114. The figure in the *zij* was 6,600 farsakhs, which would be roughly 40,000–48,000 kilometres (25,000–30,000 mi.).

9 Ṣāʿid ibn Aḥmad Andalusī, *Science in the Medieval World*, p. 46.

10 Gutas, *Greek Thought, Arabic Culture*, p. 54.

11 Ibid., p. 56.

12 Ibid., p. 107.

13 Ibid., p. 145.

14 Haig Khatchadourian, Nicholas Rescher and Ya'qub ibn Ishaq al-Kindi, 'Al-Kindi's Epistle on the Concentric Structure of the Universe', *Isis*, LVI/2 (1965), pp. 190–95.

15 George Saliba, *Islamic Science and the Making of the European Renaissance* (Cambridge, MA, 2007), p. 131.

16 Ibid., p. 73.

17 Tom Holland, *In the Shadow of the Sword* (London, 2012), p. 310.

18 The Koran, trans. N. J. Dawood (Harmondsworth, 1999), p. 399 (67.3).

19 Mohammad Ali Tabataba'i and Saida Mirsadri, 'The Qur'ānic Cosmology, as an Identity in Itself', *Arabica*, LXIII/3–4 (2016), pp. 207–9.

20 The Koran, p. 365 (50.7).

21 Tabataba'i and Mirsadri, 'The Qur'ānic Cosmology, as an Identity in Itself', p. 211.

22 Damien Janos, 'Qur'ānic Cosmography in Its Historical Perspective: Some Notes on the Formation of a Religious Worldview', *Religion*, XLII/2 (2012), p. 216.

23 The Koran, p. 212 (18.86).

24 Ibid., p. 229 (21.33).

25 Tabataba'i and Mirsadri, 'The Qur'ānic Cosmology, as an Identity in Itself', p. 217.

26 Jonathan A. C. Brown, *Misquoting Muhammad: The Challenges and Choices of Interpreting the Prophet's Legacy* (Oxford, 2014), p. 39.

27 Morrissey, *A Short History of Islamic Thought*, p. 59.

28 Muhammad al-Bukhari, *The Translation and Meanings of Sahîh Al-Bukhâri*, trans. Mohammad Muhsin Khan (Riyadh, 1997), vol. III, p. 367 (2454).

29 Muhammad ibn Yarir al-Tabari, *The History of Al-Tabari* (New York, 1989), p. 208.

30 Morrissey, *A Short History of Islamic Thought*, p. 142.

31 Anton M. Heinen, *Islamic Cosmology* (Beirut, 1982), pp. 138ff.

32 Janos, 'Qur'ānic Cosmography in Its Historical Perspective', p. 220.

33 The Koran, p. 425 (88.20).
34 Jalāl al-Dīn al-Maḥallī and Jalāl al-Dīn al-Suyūṭī, *Tafsīr Al-Jalālayn*, trans. Feras Hamza (Amman, 2007), p. 744.
35 Al-Ghazali, *Incoherence of the Philosophers*, trans. Sabih Ahmad Kamali (Lahore, 1963), p. 180.
36 Ahmad S. Dallal, *Islam, Science, and the Challenge of History* (New Haven, CT, 2010), p. 125.
37 Al-Ghazali, *The Confessions of Al-Ghazali*, ed. Claude Field (London, 1909), p. 30.
38 Brown, *Misquoting Muhammad*, p. 81.
39 Youssef M. Ibrahim, 'Muslim Edict Takes on New Force', *New York Times*, 12 February 1995.
40 Robert Lacey, *Inside the Kingdom* (London, 2009), p. 88.
41 The result can be a bit unexpected – for instance, Muslims in Los Angeles should pray facing northeast, even though Mecca is fractionally further south than California. This is for the same reason that aeroplanes flying from Europe to Los Angeles pass over northern Canada: the shortest route from two points of the Earth's surface will avoid the great bulge around the equator.
42 David A. King, 'The Sacred Geography of Islam', in *Mathematics and the Divine: A Historical Study*, ed. T. Koetsier and L. Bergmans (Amsterdam, 2005), p. 175.
43 David A. King, 'The Sacred Direction of Mecca: A Study of the Interaction of Religion and Science in the Middle Ages', *Interdisciplinary Science Reviews*, X/4 (1985), p. 321.
44 David A. King and Richard P. Lorch, 'Qibla Charts, Qibla Maps, and Related Instruments', in *Cartography in Traditional Islamic and South Asian Societies*, ed. David Woodward and J. B. Harley (Chicago, IL, 1992), p. 196.
45 Gerald R. Tibbetts, 'The Beginnings of a Cartographic Tradition', in *Cartography in Traditional Islamic and South Asian Societies*, ed. Woodward and Harley, p. 97.
46 King and Lorch, 'Qibla Charts, Qibla Maps, and Related Instruments', p. 196.
47 Ibid., p. 195.

16 Later Judaism: 'The wise men of the nations have defeated the wise men of Israel'

1 Rabbi Eliezar, *Pirke De Rabbi Eliezer*, ed. Gerald Friedlander (London, 1916), p. 28.
2 Martin Goodman, *A History of Judaism* (London, 2019), p. 327.
3 Henry Malter, *Saadia Gaon: His Life and Works* (Philadelphia, PA, 1921), p. 119.
4 Goodman, *A History of Judaism*, p. 328.
5 Saadia ben Joseph, *Rabbi Saadiah Gaon's Commentary on the Book of Creation*, trans. Michael Linetsky (Northvale, NJ, 2002), p. 57.

6 Malter, *Saadia Gaon*, p. 184.
7 Saadia ben Joseph, *Commentaire sur Le Séfer Yesira: Ou Livre de la Création* (Paris, 1891), p. 73.
8 Taro Mimura, 'The Arabic Original of (Ps.) Māshā'Allāh's Liber De Orbe: Its Date and Authorship', *British Journal for the History of Science*, XLVIII/2 (2015), p. 352.
9 Moses Maimonides, 'Foundation of the Torah', www.sefaria.org (1927), 3.2–5.
10 Moses Maimonides, *The Guide for the Perplexed*, trans. M. Friedländer (New York, 1956), p. 168 (2.11).
11 Ibid., p. 163 (2.8).
12 Ibid.
13 Ibid., p. 278 (3.14).
14 Ibid., p. 164 (2.9).
15 Natan Slifkin, *The Sun's Path at Night: The Revolution in Rabbinic Perspective on the Ptolemaic Revolution*, www.zootorah.com (2010), p. 19.
16 Goodman, *A History of Judaism*, p. 337.
17 *The Zohar*, trans. Daniel Chanan Matt (Stanford, CA, 2004), vol. VII, pp. 48 and 51.
18 Shulamit Laderman, *Images of Cosmology in Jewish and Byzantine Art* (Leiden, 2013), p. 102.
19 Slifkin, *The Sun's Path at Night*, p. 13.
20 Ibid., p. 43.

17 Europe in the Early Middle Ages: 'Equally round in all directions'

1 Ambrose of Milan, *Hexameron, Paradise, and Cain and Abel*, trans. John J. Savage (Washington, DC, 1961), p. 21 (1.22).
2 Lactantius, *The Divine Institutes*, ed. Anthony Bowen and Peter Garnsey (Liverpool, 2003), p. 213 (3.24).
3 Leo C. Ferrari, 'Astronomy and Augustine's Break with the Manichees', *Revue d'Etudes Augustiniennes et Patristiques*, XIX/3–4 (1973), p. 272.
4 Augustine of Hippo, *Confessions*, trans. R. S. Pine-Coffin (Harmondsworth, 1961), p. 93 (5.3).
5 Ferrari, 'Astronomy and Augustine's Break with the Manichees', p. 274.
6 Augustine of Hippo, *Confessions*, p. 98 (5.7).
7 Henry Chadwick, *Augustine of Hippo* (Oxford, 2010), p. 37.
8 Augustine of Hippo, *On Genesis: A Refutation of the Manichees, Unfinished Literal Commentary on Genesis, the Literal Meaning of Genesis*, ed. Edmund Hill (Hyde Park, NY, 2002), p. 186 (1.19).
9 Ibid., pp. 201, 210 (2.9, 2.15).
10 A dissenter from this scholarly consensus was an Australian expert on Augustinian theology, based in New Brunswick, Canada, called

Leo Ferrari. It's hard to know how we should take his contention that Augustine was a flat-earther because Ferrari was long-time president of the Flat Earth Society of Canada. This distinguished body appears to have been an elaborate practical joke (motto: 'We're on the level') cooked up by Ferrari and his mates over a few too many drinks. It lasted for years before becoming so infamous that it had to be closed down in case people started to take it seriously. See Leo Ferrari, 'Augustine's Cosmography', *Augustinian Studies*, XXVII/2 (1996), pp. 129–77; Christine Garwood, *Flat Earth: The History of an Infamous Idea* (London, 2007), p. 280.

11 Cassiodorus, *Institutions of Divine and Secular Learning*, trans. James W. Halprin and Mark Vessey (Liverpool, 2004), pp. 225, 227 (2.6.4, 2.7.3).

12 Isidore of Seville, *The Etymologies of Isidore of Seville*, trans. Stephan A. Barney (Cambridge, 2006), p. 285 (14.2.1).

13 W. M. Stevens, 'The Figure of the Earth in Isidore's "De Natura Rerum"', *Isis*, LXXI/2 (1980), pp. 268–77; William D. McCready, 'Isidore, the Antipodeans, and the Shape of the Earth', *Isis*, LXXXVII/1 (1996), pp. 108–27; Andrew Fear, 'Putting the Pieces Back Together: Isidore and De Natura Rerum', in *Isidore of Seville and His Reception in the Early Middle Ages*, ed. Andrew Fear and Jamie Wood (Amsterdam, 2016), pp. 75–92 (p. 76).

14 Fear, 'Putting the Pieces Back Together', p. 85.

15 Isidore of Seville, *On the Nature of Things*, trans. Calvin B. Kendall and Faith Wallis (Liverpool, 2016), p. 144 (20).

16 Fear, 'Putting the Pieces Back Together', p. 79.

17 Isidore of Seville, *On the Nature of Things*, p. 258.

18 Ibid., p. 129 (10.3).

19 Ibid., p. 172 (44.2).

20 Ibid., p. xxi.

21 Michael W. Herren, ed. and trans., *The Cosmography of Aethicus Ister: Edition, Translation, and Commentary* (Turnhout, 2011), p. xxxvi. Michael Herren, the *Cosmography*'s modern translator, suggests that its eccentric world picture poked fun at Cosmas Indicopleustes. In my view, that the author of the *Cosmography* pretended to adopt a flat-earth cosmology, even though he knew the Earth was really a sphere, stretches credibility to breaking point. And, while it is just about possible that he knew of Cosmas' book (despite it being in Greek), it's highly unlikely that there was anyone of his acquaintance who would have got the joke.

22 Ibid., p. 25.

23 Marina Smyth, *Understanding the Universe in Seventh-Century Ireland* (Woodbridge, 1996), p. 278.

24 The anonymous author of *On the Arrangement of Created Things*, a seventh-century Irish work long attributed to Isidore himself, is a case in point. The Irish monk responsible was sensible enough

not to rule out possibilities about the shape of the universe, as long as they were consistent with the Bible. Like Augustine, he was equally happy with the heavens covering the Earth like a disc, or being shaped like an egg and surrounding the whole of creation (see Marina Smyth, 'The Seventh-Century Hiberno-Latin Treatise "Liber De Ordine Creaturarum", a Translation', *Journal of Medieval Latin*, XXI (2011), p. 172). When it came to the Earth itself, he described it as an *orbis* divided into four segments to the north, south, east and west (Smyth, 'Liber De Ordine Creaturarum', p. 195), which implies a disc rather than a sphere.

25 Bede, *The Reckoning of Time*, trans. Faith Wallis (Liverpool, 1999), p. 78 (27).

26 Ibid., p. 91 (32).

27 Bede, *Ecclesiastical History of the English People*, trans. Leo Sherley-Price (Harmondsworth, 1990), p. 359.

28 Bruce Eastwood, *Ordering the Heavens: Roman Astronomy and Cosmology in the Carolingian Renaissance* (Leiden, 2007), pp. 88 and 127.

29 Ibid., p. 9.

30 Stephen C. McCluskey, *Astronomies and Cultures in Early Medieval Europe* (Cambridge, 1998), p. 133.

31 Cicero, 'Academica', in *On the Nature of the Gods. Academics*, trans. H. Rackham (Cambridge, MA, 1933), p. 627 (2.123).

32 Pliny the Elder, *Natural History*, trans. H. Rackham (Cambridge, MA, 1938), vol. I, p. 297 (2.65).

33 Macrobius, *Commentary on the Dream of Scipio*, trans. William Harris Stahl (New York, 1952), p. 204 (2.5.25).

34 Lactantius, *The Divine Institutes*, ed. Anthony Bowen and Peter Garnsey (Liverpool, 2003), p. 213 (3.24).

35 Augustine of Hippo, *City of God*, trans. Henry Bettinson (Harmondsworth, 1984), p. 664 (16.9).

36 Isidore of Seville, *The Etymologies*, p. 199 (9.2.133).

37 Bede, *The Reckoning of Time*, p. 99 (34).

38 John Carey, 'Ireland and the Antipodes: The Heterodoxy of Virgil of Salzburg', *Speculum*, LXIV/1 (1989), p. 1. Among the Irish, legends about the underworld became entangled with classical concepts of the Globe and the antipodes. According to one account, the Sun illuminated the lands beneath the Earth as it passed them on its way to rising over the Northern Hemisphere. As an eleventh-century Irish poet wrote, while the Sun was over the antipodes, 'it shines on the hosts of the youths in the playing fields, who cry out to heaven for fear of the beast'. (See Carey, 'Ireland and the Antipodes', p. 5.) Any British athlete who has competed in Australia will understand the sentiment.

18 High Medieval Views of the World: 'The Earth has
the shape of a globe'

1 Taro Mimura, 'The Arabic Original of (Ps.) Māshā'Allāh's Liber
De Orbe: Its Date and Authorship', British Journal for the History of
Science, XLVIII/2 (2015), pp. 321–52.

2 Ibid., p. 352.

3 James Hannam, God's Philosophers: How the Medieval World Laid
the Foundations of Modern Science (London, 2009), p. 69.

4 Edward Grant, ed., A Source Book in Medieval Science (Cambridge,
MA, 1974), pp. 36–8.

5 Barbara Obrist, 'William of Conches, Māshā'Allāh, and Twelfth-
Century Cosmology', Archives d'histoire doctrinale et littéraire du
Moyen Âge, LXXVI/1 (2009), p. 56.

6 William of Conches, A Dialogue on Natural Philosophy:
Dragmaticon Philosophiae, trans. Italo Ronca and Matthew Curr
(Notre Dame, IN, 1997), p. 4 (1.1.5).

7 Ibid., p. 121 (6.2.2).

8 There were two versions of the Book of the Orb in circulation, one
with 27 chapters and one with 40. William evidently used the
longer one. See Obrist, 'William of Conches, Māshā'Allāh, and
Twelfth-Century Cosmology', p. 33.

9 Ibid., p. 58.

10 William of Conches, A Dialogue on Natural Philosophy, pp. 121, 122
(6.2.2, 6.2.7).

11 Eric M. Ramírez-Weaver, 'William of Conches, Philosophical
Continuous Narration, and the Limited Worlds of Medieval
Diagrams', Studies in Iconography, XXX (2009), p. 6.

12 Hannam, God's Philosophers, p. 74.

13 Seb Falk, The Light Ages: A Medieval Journey of Discovery (London,
2020), p. 86.

14 Grant, A Source Book in Medieval Science, p. 444.

15 Jill Tattersall, 'Sphere or Disc? Allusions to the Shape of the Earth
in Some Twelfth-Century and Thirteenth-Century Vernacular
French Works', Modern Language Review, LXXVI/1 (1981), p. 43.

16 Ibid., p. 34. The idiom of Old French, like modern English and
ancient Latin, was to refer to the Earth as 'round' rather than
'spherical', so the poets felt it necessary to clarify that they meant a
sphere and not a disc.

17 There was a sect, the Cathars, whose doctrines were similar
to Manichaeans, active in southern France at the time, but it is
unlikely Dante cared about their theology.

18 The Travels of Sir John Mandeville, trans. C.W.R.D Moseley
(Harmondsworth, 1983), p. 127.

19 Ibid., p. 130.

20 Ibid., p. 129.
21 While we call the item of regalia an 'orb' in English, its Latin name is *Globus cruciger*, which avoids the ambiguity of the word *orbis*.
22 David Woodward, 'Medieval *Mappaemundi*', in *Cartography in Prehistoric, Ancient, and Medieval Europe and the Mediterranean*, ed. David Woodward and J. B. Harley (Chicago, IL, 1987), p. 301.
23 David Wootton, *The Invention of Science: A New History of the Scientific Revolution* (London, 2015), p. 115.
24 Aristotle, 'Meteorology', in *The Complete Works of Aristotle*, ed. Jonathan Barnes (Princeton, NJ, 1984), p. 587 (362b).
25 Alessandro Scafi, 'Defining Mappamundi', in *The Hereford World Map: Medieval World Maps and Their Context*, ed. P.D.A. Harvey (London, 2006), p. 346.
26 Felipe Fernández-Armesto, *Pathfinders: A Global History of Exploration* (New York, 2006), p. 81.
27 Woodward, 'Medieval *Mappaemundi*'.
28 J. Williams, 'Isidore, Orosius and the Beatus Map', *Imago Mundi*, XLIX (1997), p. 17.
29 Ibid., p. 293 (14.5.17).

19 Columbus and Copernicus: 'New worlds will be found'

1 Washington Irving, *Life and Voyages of Christopher Columbus* (London, 1909), p. 60.
2 Ptolemy of Alexandria, *Ptolemy's Geography: An Annotated Translation of the Theoretical Chapters*, trans. J. L. Berggren and Alexander Jones (Princeton, NJ, 2000), p. 43.
3 Ibid., p. 21.
4 W.G.L. Randles, 'The Evaluation of Columbus' "India" Project by Portuguese and Spanish Cosmographers in the Light of the Geographical Science of the Period', *Imago Mundi*, XLII (1990), p. 54. Toscanelli based his contention on the views of a certain ancient Greek called Marinus, whom Ptolemy had criticized in the *Geography*. See Ptolemy of Alexandria, *Ptolemy's Geography*, p. 71.
5 Christopher Columbus, *The Four Voyages*, ed. J. M. Cohen (Harmondsworth, 1969), p. 13.
6 Randles, 'The Evaluation of Columbus' "India" Project', p. 52.
7 There is mention of a terrestrial globe constructed 'according to Ptolemy's description' from 1443, but the object does not survive. See David Wootton, *The Invention of Science: A New History of the Scientific Revolution* (London, 2015), p. 121.
8 Randles, 'The Evaluation of Columbus' "India" Project', p. 61.
9 J. H. Elliott, *The Old World and the New* (Cambridge, 1970), p. 29.
10 David N. Livingstone, *The Geographical Tradition: Episodes in the History of a Contested Enterprise* (Oxford, 1992), p. 34.
11 Seneca the Younger, 'Medea', in *Tragedies*, ed. John G. Fitch (Cambridge, MA, 2018), vol. I, p. 379 (ll. 375–9).

12 Columbus, *The Four Voyages*, p. 218.
13 See Chapter Seventeen and following in James Hannam, *God's Philosophers: How the Medieval World Laid the Foundations of Modern Science* (London, 2009).
14 Nicolaus Copernicus, *On the Revolutions of the Heavenly Spheres*, ed. Charles Glenn Wallis (Amherst, MA, 1995), pp. 9–11.
15 Maurice A. Finocchiaro, *Retrying Galileo, 1633–1992* (Berkeley, CA, 2005), p. 21.
16 Copernicus, *On the Revolutions of the Heavenly Spheres*, p. 19.
17 Ibid., p. 10.

20 China: 'The heavens are round and the Earth is square'

1 Valerie Hansen, *The Open Empire: A History of China to 1600* (New York, 2000), p. 126.
2 'Q' denotes a sound something like a 'Ch' in the approved pinyin system of the modern Chinese government, while 'Zh' is pronounced a bit like 'J'.
3 Christopher Cullen, *Heavenly Numbers: Astronomy and Authority in Early Imperial China* (Oxford, 2017), p. 22.
4 Ibid., p. 31.
5 Ibid., p. 393.
6 Liu An, *The Huainanzi: A Guide to the Theory and Practice of Government in Early Han China*, ed. John Major et al., trans. John Major (New York, 2010), p. 115.
7 Ibid., p. 279.
8 Roel Sterckx, *Chinese Thought: From Confucius to Cook Ding* (London, 2019), p. 350.
9 Cullen, *Heavenly Numbers*, p. 203.
10 *The Classic of the Mountains and the Sea*, trans. Anne Birrell (Harmondsworth, 1999).
11 Liu An, *The Huainanzi*, pp. 144, 157.
12 Hansen, *The Open Empire*, p. 91.
13 John B. Henderson, *The Development and Decline of Chinese Cosmology* (New York, 1984), p. 72.
14 Qiong Zhang, *Making the New World Their Own: Chinese Encounters with Jesuit Science in the Age of Discovery* (Leiden, 2015), p. 27.
15 Cullen, *Heavenly Numbers*, p. 208.
16 Christopher Cullen, *Astronomy and Mathematics in Ancient China: The Zhou Bi Suan Jing* (Cambridge, 1996), p. 178.
17 Ibid., p. 189.
18 Ibid., p. 180.
19 Cullen, *Heavenly Numbers*, pp. 276–8.
20 Ibid., p. 280.
21 Ibid., p. 283.
22 Joseph Needham and Ling Wang, *Mathematics and the Sciences of the Heavens and the Earth* (Cambridge, 1959), p. 218.

23 Jin Zumeng, 'A Critique of "Zhang Heng's Theory of a Spherical Earth"', in *Chinese Studies in the History and Philosophy of Science and Technology*, ed. Fan Dainian and Robert S. Cohen (Dordrecht, 1996), p. 432.

24 Christopher Cullen, 'Joseph Needham on Chinese Astronomy', *Past and Present*, LXXXVII (1980), p. 42.

25 Jeffrey Kotyk, 'The Chinese Buddhist Approach to Science: The Case of Astronomy and Calendars', *Journal of Dharma Studies*, III/2 (2020), p. 279.

26 Yuzhen Guan, 'A New Interpretation of Shen Kuo's *Ying Biao Yi*', *Archive for History of Exact Sciences*, LXIV (2010), p. 712.

27 Jeffrey Kotyk, 'Examining Amoghavajra's Flat-Earth Cosmology: Religious vs Scientific Worldviews in Buddhist Astrology', *Studies in Chinese Religions*, VII/2 (2021), p. 205.

28 Kotyk, 'The Chinese Buddhist Approach to Science', p. 284.

29 Bill M. Mak, 'Yusi Jing – A Treatise of "Western" Astral Science in Chinese and Its Versified Version Xitian Yusi Jing', SCIAMVS, XV (2014), p. 128.

30 Jan Vrhovski, 'Apologeticism in Chinese Nestorian Documents from the Tang Dynasty', *Asian Studies I*, XXVII/2 (2013), p. 58.

31 Hansen, *The Open Empire*, p. 242.

32 Ibid., p. 264.

33 Ibid., p. 412.

34 Ibid., p. 270.

35 Qiao Yang, 'From the West to the East, from the Sky to the Earth: A Biography of Jamāl Al-Dīn', *Asiatische Studien – Études Asiatiques*, LXXI/4 (2018), p. 1234.

36 Ho Peng Yoke, *Li, Qi and Shu: An Introduction to Science and Civilization in China* (Seattle, WA, 1987), p. 167.

37 Yang, 'From the West to the East', p. 1235.

38 Ibid., p. 1238.

39 Ho Peng Yoke, *Li, Qi and Shu*, p. 167.

40 Louise Levathes, *When China Ruled the Seas: The Treasure Fleet of the Dragon Throne, 1405–1433* (Oxford, 1994), p. 119.

41 Ibid., p. 20.

42 Song Zhenghai and Chen Chuankang, 'Why Did Zheng He's Sea Voyage Fail to Lead the Chinese to Make the "Great Geographic Discovery"?', in *Chinese Studies in the History and Philosophy of Science and Technology: Boston Studies in the Philosophy of Science*, ed. Fan Dainian and Robert S. Cohen (Dordrecht, 1996), p. 308.

43 Zhang, *Making the New World Their Own*, p. 61.

44 Sterckx, *Chinese Thought: From Confucius to Cook Ding*, p. 4.

21 China and the West: 'Like the yolk in a hen's egg'

1 Mary Laven, *Mission to China* (London, 2011), p. 22.

2 Joseph Needham and Ling Wang, *Mathematics and the Sciences of the Heavens and the Earth* (Cambridge, 1959), p. 438.
3 Qiong Zhang, *Making the New World Their Own: Chinese Encounters with Jesuit Science in the Age of Discovery* (Leiden, 2015), p. 71.
4 Ibid., p. 47.
5 Ibid., p. 57.
6 Laven, *Mission to China*, p. 224.
7 Roel Sterckx, *Chinese Thought: From Confucius to Cook Ding* (London, 2019), p. 320.
8 Laven, *Mission to China*, p. 223.
9 Benjamin A. Elman, *On Their Own Terms: Science in China, 1550–1900* (Cambridge, MA, 2005), p. 94.
10 Liam Matthew Brockey, *Journey to the East: The Jesuit Mission to China, 1579–1724* (Cambridge, MA, 2007), p. 68.
11 Elman, *On Their Own Terms*, p. 93.
12 Toby E. Huff, *Intellectual Curiosity and the Scientific Revolution: A Global Perspective* (Cambridge, 2011), p. 90.
13 Elman, *On Their Own Terms*, p. 97.
14 Ibid., p. 98.
15 Ibid., p. 134.
16 Ibid., p. 135.
17 Zhang, *Making the New World Their Own*, p. 156.
18 Ibid., p. 137.
19 Brockey, *Journey to the East*, p. 128.
20 Elman, *On Their Own Terms*, p. 142.
21 Ibid., p. 144.
22 Pingyi Chu, 'Scientific Dispute in the Imperial Court: The 1664 Calendar Case', *Chinese Science*, XIV (1997), p. 34.
23 Brockey, *Journey to the East*, p. 200.
24 Elman, *On Their Own Terms*, p. 168.
25 Needham and Wang, *Mathematics and the Sciences of the Heavens and the Earth*, p. 499.
26 Elman, *On Their Own Terms*, p. 262.
27 John B. Henderson, *The Development and Decline of Chinese Cosmology* (New York, 1984), p. 237.
28 Elman, *On Their Own Terms*, p. 267.
29 Maurice A. Finocchiaro, *Retrying Galileo, 1633–1992* (Berkeley, CA, 2005), p. 139.
30 Benjamin A. Elman, *A Cultural History of Modern Science in China* (Cambridge, MA, 2006), p. 43.
31 Zhang, *Making the New World Their Own*, p. 200.
32 Ibid., p. 198.
33 John B. Henderson, 'Ch'ing Scholars' Views of Western Astronomy', *Harvard Journal of Asiatic Studies*, XLVI/1 (1986), p. 141.

22 The Globe Goes Global: 'At the round Earth's imagined corners'

1 This provides a precise figure if Polaris is exactly stationary over the North Pole. This isn't quite true. Polaris is currently very close to being on a bearing of true north, but it still moves around in the sky as the Earth turns. It is usually a distance equal to a full moon away from the pole.

2 Maria M. Portuondo, 'Lunar Eclipses, Longitude and the New World', *Journal of the History of Astronomy*, XL/3 (1990), p. 251.

3 Larrie D. Ferreiro, *Measure of the Earth* (New York, 2011), p. 6.

4 Ibid., p. 8; Isaac Newton, 'Principia', in *On the Shoulders of Giants: The Great Works of Physics and Astronomy*, ed. Stephen Hawking (Philadelphia, PA, 2002), p. 1066.

5 Felipe Fernández-Armesto, *Pathfinders: A Global History of Exploration* (New York, 2006), p. 301.

6 Bruce Trigger, *Understanding Early Civilisations: A Comparative Study* (Cambridge, 2003), p. 447.

7 Ibid., p. 448.

8 Nicholas Campion, *Astrology and Cosmology in the World's Religions* (New York, 2012), p. 29.

9 Ibid., p. 36.

10 Edward Burnett Tylor, *Anthropology* (London, 1913), p. 333.

11 Ibid.

12 T. B. Macaulay, 'Macaulay's Minute', in *Selections from Educational Records, Part I (1781–1839)*, ed. H. Sharp (Delhi, 1965), p. 111.

13 John Donne, *Selected Poems* (London, 2006), p. 180.

14 Donald S. Lopez Jr, *Buddhism and Science: A Guide for the Perplexed* (Chicago, IL, 2008), p. 42.

15 See Chapter Fourteen of R. S. Sharma, *India's Ancient Past* (New Delhi, 2005).

16 Jeffrey Kotyk, 'Celestial Deities in the Flat-Earth Buddhist Cosmos and Astrology', in *Intersections of Religion and Astronomy*, ed. Chris Corbally, Darry Dinell and Aaron Ricker (London, 2020), p. 37.

17 Jeffrey Kotyk, 'The Chinese Buddhist Approach to Science: The Case of Astronomy and Calendars', *Journal of Dharma Studies*, III/2 (2020), p. 4.

18 Ibid., p. 46.

19 Ibid., p. 51.

20 Ibid., p. 57.

21 Ibid., p. 69.

22 Lopez, *Buddhism and Science*, p. 54.

23 Christine Garwood, *Flat Earth: The History of an Infamous Idea* (London, 2007), p. 133.

24 Matthew 7:7.

25 Samuel Birley Rowbotham, *Zetetic Astronomy: Earth Not a Globe* (London, 1881), p. 354.
26 Ibid.
27 Garwood, *Flat Earth*, p. 70.
28 Ibid., p. 105.
29 Ibid., p. 141.
30 Ibid., p. 152.
31 Ibid., p. 166.
32 Ibid., p. 177.

23 Today: 'There's nothing particularly exciting about a round world'

1 Christine Garwood, *Flat Earth: The History of an Infamous Idea* (London, 2007), p. 193.
2 Ibid., p. 211.
3 Ibid., p. 192.
4 Ibid., p. 217.
5 Ibid., p. 254.
6 The author's Australian wife assures him this is no longer the case; Garwood, *Flat Earth*, p. 323.
7 Ibid., p. 348.
8 Marco Silva, 'Flat Earth: How Did YouTube Help Spread a Conspiracy Theory?', www.bbc.co.uk, 18 July 2019.
9 '"Mad" Mike Hughes Dies after Crash-Landing Homemade Rocket', www.bbc.co.uk, 23 February 2020.
10 Michael Newton Keas, *Unbelievable: 7 Myths About the History and Future of Science and Religion* (Wilmington, DE, 2019), p. 47.
11 J. H. Parry, *The Age of Reconnaissance: Discovery, Exploration and Settlement 1450 to 1650* (London, 1963), p. 10.
12 Daniel J. Boorstin, *The Discoverers: A History of Man's Search to Know His World and Himself* (New York, 1983).
13 Jeffrey Burton Russell, *Inventing the Flat Earth: Columbus and Modern Historians* (Westport, VA, 1991).
14 Umberto Eco, *Serendipities: Language and Lunacy* (London, 1999).
15 John Henry, *Knowledge Is Power: How Magic, the Government and an Apocalyptic Vision Inspired Francis Bacon to Create Modern Science* (Cambridge, 2002), p. 85.
16 Thomas Paine, 'The Age of Reason Part One', in *The Thomas Paine Reader*, ed. Michael Foot (Harmondsworth, 1987), p. 431.
17 Thomas Jefferson, *Notes on the State of Virginia* (Boston, MA, 1832), p. 167.
18 Russell, *Inventing the Flat Earth*, p. 58.
19 Andrew Dickson White, *A History of the Warfare of Science with Theology in Christendom*, 2 vols (New York, 1896), vol. I, p. 97.
20 David S. Cohen, *The Simpsons*, 'Lisa the Skeptic', Series 8, Episode 9 (1997).

21 Stephen Jay Gould, *Rocks of Ages: Science and Religion in the Fullness of Life* (London, 2001).

22 Ekmeleddin İhsanoğlu, 'Modern Islam', in *Science and Religion around the World*, ed. John Hedley Brooke and Ronald L. Numbers (New York, 2011), p. 165.

23 David Hutchings and James C. Ungureanu, *Of Popes and Unicorns: Science, Christianity, and How the Conflict Thesis Fooled the World* (Oxford, 2021), p. 176.

24 Roger Lancelyn Green and Walter Hooper, *C. S. Lewis: A Biography* (London, 2002), p. 80.

25 C. S. Lewis, *The Discarded Image: An Introduction to Medieval and Renaissance Literature* (Cambridge, 1994), p. 140.

26 Ironically, Pauline Baynes (1922–2008), who drew the illustrations for the book, placed a globe in the captain's cabin of the *Dawn Treader*. See C. S. Lewis, *The Voyage of the Dawn Treader* (London, 1980), p. 24.

27 Lewis, *The Voyage of the Dawn Treader*, p. 176.

28 J.R.R Tolkien, *The Lost Road and Other Writings* (London, 1987), p. 16.

29 J.R.R. Tolkien, *The Monsters and the Critics and Other Essays* (London, 1997), p. 18.

30 J.R.R. Tolkien, *Beowulf: A Translation and Commentary* (London, 2014), p. 225.

31 John Garth, *The Worlds of J.R.R. Tolkien* (London, 2020), p. 40.

32 Lewis, *The Discarded Image*, p. 141.

33 J.R.R. Tolkien, *Tolkien on Fairy-Stories*, ed. Verlyn Flieger and Douglas A. Anderson (London, 2014), p. 61.

34 John Rateliff, 'The Flat Earth Made Round and Tolkien's Failure to Finish the Silmarillion', *Journal of Tolkien Research*, IX/1 (2020), p. 13.

35 Terry Pratchett, *The Colour of Magic* [1983] (London, 1985), p. 207.

36 Pratchett had already used it in a science fiction novel, *Strata*, published a couple of years before *The Colour of Magic*.

Afterword

1 Nicole Oresme, a fourteenth-century master at the University of Paris, was the first to convincingly do this. See James Hannam, *God's Philosophers: How the Medieval World Laid the Foundations of Modern Science* (London, 2009), p. 187.

2 Anthony Kenny, *A New History of Western Philosophy* (Oxford, 2010), p. 295.

BIBLIOGRAPHY

Ambrose of Milan, *Hexameron, Paradise, and Cain and Abel*, ed. John J.
    Savage (Washington, DC, 1961)
Andalusī, Ṣāʿid ibn Aḥmad, *Science in the Medieval World: Book of
    the Categories of Nations*, ed. Semaʿan I. Salem and Alok Kumar
    (Austin, TX, 1991)
Aristophanes, 'The Clouds', in *Lysistrata and Other Plays*, ed. Alan
    Sommerstein (Harmondsworth, 1973), pp. 105–74
Aristotle, *Politics*, ed. T. A. Sinclair and Trevor J. Saunders
    (Harmondsworth, 1981)
—, 'Meteorology', in *The Complete Works of Aristotle*, ed. Jonathan
    Barnes (Princeton, NJ, 1984), pp. 555–625
—, 'On Generation and Corruption', in *The Complete Works of
    Aristotle*, ed. Jonathan Barnes (Princeton, NJ, 1984), pp. 512–54
—, 'On the Heavens', in *The Complete Works of Aristotle*, ed. Jonathan
    Barnes (Princeton, NJ, 1984), pp. 447–511
—, *Metaphysics*, ed. Hugh Lawson-Tancred (Harmondsworth, 1998)
Arrian, *The Campaigns of Alexander*, ed. Aubrey de Sélincourt and J. R.
    Hamilton (Harmondsworth, 1971)
Aryabhata, *The Aryabhatiya of Aryabhata*, ed. David Eugene Smith
    (Chicago, IL, 1930)
Augustine of Hippo, *Confessions*, ed. R. S. Pine-Coffin
    (Harmondsworth, 1961)
—, *City of God*, ed. Henry Bettinson (Harmondsworth, 1984)
—, *On Genesis: A Refutation of the Manichees, Unfinished Literal
    Commentary on Genesis, the Literal Meaning of Genesis*, ed. Edmund
    Hill (Hyde Park, NY, 2002)
Bagnall, Roger S., 'Alexandria: Library of Dreams', *Proceedings of the
    American Philological Society*, CXLVI/4 (2002), pp. 348–62
Bakker, Frederik A., *Epicurean Meteorology: Sources, Method, Scope and
    Organization* (Leiden, 2016)

—, 'Vergilius Astronomiae Ignarus? A Vindication of Virgil's
   Astronomical Knowledge in *Georgics* 1.231–258', *Mnemosyne*,
   LXXII/4 (2019), pp. 621–46
Barnes, Jonathan, *Aristotle* (Oxford, 1982)
Basil of Caesarea, 'Hexaemeron', in *St Basil: Letters and Select Works*,
   ed. Philip Schaff and Henry Wace (New York, 1895), pp. 51–108
Bede, *Ecclesiastical History of the English People*, ed. Leo Sherley-Price
   (Harmondsworth, 1990)
—, *The Reckoning of Time*, ed. Faith Wallis (Liverpool, 1999)
—, *On the Nature of Things and on Times*, ed. Faith Wallis (Liverpool,
   2010)
Bigwood, J. M., 'Aristotle and the Elephant Again', *American Journal of
   Philology*, CXIV/4 (1993), pp. 537–55
Birrell, Anne, ed., *The Classic of the Mountains and the Sea*
   (Harmondsworth, 1999)
Blacker, Carmen, Michael Loewe and J. Martin Plumley, eds, *Ancient
   Cosmologies* (London, 1975)
Bose, Devabrata M., Samarendra Nath Sen and Bidare V.
   Subbarayappa, *A Concise History of Science in India* (New Delhi,
   1971)
Bowles, Adam, trans., *The Mahabharata* VIII (New York, 2006)
Brockey, Liam Matthew, *Journey to the East: The Jesuit Mission to China,
   1579–1724* (Cambridge, MA, 2007)
Brown, Jonathan A. C., *Misquoting Muhammad: The Challenges and
   Choices of Interpreting the Prophet's Legacy* (Oxford, 2014)
al-Bukhari, Muhammad, *The Translation and Meanings of Sahîh
   Al-Bukhâri*, ed. Mohammad Muhsin Khan (Riyadh, 1997)
Burch, George Bosworth, 'The Counter-Earth', *Osiris*, XI (1954),
   pp. 267–94
Burkert, Walter, *Lore and Science in Ancient Pythagoreanism*
   (Cambridge, MA, 1972)
Cameron, Alan, 'The Last Days of the Academy at Athens', *Proceedings
   of the Cambridge Philological Society*, XV (1969), pp. 7–29
Campion, Nicholas, *Astrology and Cosmology in the World's Religions*
   (New York, 2012)
Carey, John, 'Ireland and the Antipodes: The Heterodoxy of Virgil of
   Salzburg', *Speculum*, LXIV/1 (1989), pp. 1–10
Cassiodorus, *Institutions of Divine and Secular Learning*, ed. James W.
   Halporn and Mark Vessey (Liverpool, 2004)
Caudano, Anne-Laurence, 'Un Univers sphérique ou voûté?
   Survivance de la cosmologie Antiochienne à Byzance (XIe et XIIe
   S.)', *Byzantion*, LXXVIII (2008), pp. 66–86
—, '"Le Ciel a la forme d'un cube ou a été dressé comme une peau":
   Pierre le philosophe et l'orthodoxie du savoir astronomique sous
   Manuel Ier Comnène', *Byzantion*, LXXXI (2011), pp. 19–73
Chadwick, Henry, *Augustine of Hippo* (Oxford, 2010)

Chrysostom, John, 'Homilies of the Epistle to the Hebrews', in *Homilies on the Gospel of St John and the Epistle to the Hebrews*, ed. Philip Schaff (New York, 1889), pp. 647–974

Chu, Pingyi, 'Scientific Dispute in the Imperial Court: The 1664 Calendar Case', *Chinese Science*, xiv (1997), pp. 7–34

Cicero, 'On the Republic', in *Cicero: On the Republic, On the Laws*, ed. Clinton W. Keyes (Cambridge, MA, 1928)

—, 'Academica', in *On the Nature of the Gods. Academics*, ed. H. Rackham (Cambridge, MA, 1933), pp. 399–659

—, *The Nature of the Gods*, ed. H.C.P. McGregor and J. M. Ross (Harmondsworth, 1972)

—, *Letters to Friends*, ed. D. R. Shackleton Bailey (Cambridge, MA, 2001)

Clagett, Marshall, *Ancient Egyptian Science: A Source Book* (Philadelphia, PA, 1995)

Clark, Travis Lee, 'Imaging the Cosmos: The Christian Topography by Kosmas Indikopleustes', PhD thesis, Temple University, 2008

Cleomedes, *Cleomedes' Lectures on Astronomy: A Translation of the Heavens*, ed. Alan C. Bowen and Robert B. Todd (Berkeley, CA, 2004)

Columbus, Christopher, *The Four Voyages*, ed. J. M. Cohen (Harmondsworth, 1969)

Copenhaver, Brian, and Charles Schmitt, *Renaissance Philosophy* (Oxford, 1992)

Copernicus, Nicolaus, *On the Revolutions of the Heavenly Spheres*, ed. Charles Glenn Wallis (Amherst, MA, 1995)

Cormack, Lesley B., 'Flat Earth or Round Sphere: Misconceptions of the Shape of the Earth and the Fifteenth-Century Transformation of the World', *Ecumene*, 1/4 (1994), pp. 363–85

Cornford, Francis, *From Religion to Philosophy: A Study in the Origins of Western Speculation* (New York, 1957)

Cottrell, Emily, and Micah Ross, 'Persian Astrology: Dorotheus and Zoroaster According to the Medieval Arabic Sources (8th–11th Century.)', in *Proceedings of the 8th European Conference of Iranian Studies* (St Petersburg, 2019), pp. 87–105

Couprie, Dirk, 'Some Remarks on the Earth in Plato's *Phaedo*', *Hyperboreus*, xi (2005), pp. 192–204

Cullen, Christopher, 'Joseph Needham on Chinese Astronomy', *Past and Present*, lxxxvii (1980), pp. 39–53

—, *Astronomy and Mathematics in Ancient China: The Zhou Bi Suan Jing* (Cambridge, 1996)

—, *Heavenly Numbers: Astronomy and Authority in Early Imperial China* (Oxford, 2017)

Cullen, Christopher, and Catherine Jami, 'Christmas 1668 and After: How Jesuit Astronomy Was Restored to Power in Beijing', *Journal for the History of Astronomy*, li/1 (2020), pp. 3–50

Cunliffe, Barry, *By Steppe, Desert, and Ocean: The Birth of Eurasia* (Oxford, 2015)

Dallal, Ahmad S., *Islam, Science, and the Challenge of History* (New Haven, CT, 2010)

Daryaee, Touraj, 'Mind, Body, and the Cosmos: Chess and Backgammon in Ancient Persia', *Iranian Studies*, XXXV/4 (2002), pp. 281–312

—, *Sasanian Persia: The Rise and Fall of an Empire* (London, 2009)

The Koran, ed. N. J. Dawood (Harmondsworth, 1999)

Dicks, D. R., *Early Greek Astronomy to Aristotle* (London, 1970)

Diogenes Laertius, *Lives of Eminent Philosophers*, ed. R. D. Hicks (Cambridge, MA, 1925)

Dodds, E. R., *The Greeks and the Irrational* (Berkeley, CA, 1951)

Eastwood, Bruce, *Ordering the Heavens: Roman Astronomy and Cosmology in the Carolingian Renaissance* (Leiden, 2007)

Elliott, J. H., *The Old World and the New* (Cambridge, 1970)

Elman, Benjamin A., *On Their Own Terms: Science in China, 1550–1900* (Cambridge, MA, 2005)

—, *A Cultural History of Modern Science in China* (Cambridge, MA, 2006)

Elweskiöld, Birgitta, 'John Philoponus against Cosmas Indicopleustes: A Christian Controversy on the Structure of the World in Sixth-Century Alexandria', PhD thesis, Lund University, 2005

Eratosthenes and Hyginus, *Constellation Myths with Aratus's Phaenomena*, ed. Robin Hard (Oxford, 2015)

Evans, James, and Christián Carlos Carman, 'The Two Earths of Eratosthenes', *Isis*, CVI/1 (2015), pp. 1–16

Falk, Seb, *The Light Ages: A Medieval Journey of Discovery* (London, 2020)

Fear, Andrew, 'Putting the Pieces Back Together: Isidore and *De Natura Rerum*', in *Isidore of Seville and His Reception in the Early Middle Ages*, ed. Andrew Fear and Jamie Wood (Amsterdam, 2016), pp. 75–92

Fear, Andrew, and Jamie Wood, eds, *Isidore of Seville and His Reception in the Early Middle Ages* (Amsterdam, 2016)

Fernández-Armesto, Felipe, *Pathfinders: A Global History of Exploration* (New York, 2006)

Ferrari, Leo, 'Astronomy and Augustine's Break with the Manichees', *Revue d'Etudes Augustiniennes et Patristiques*, XIX/3–4 (1973), pp. 263–76

—, 'Augustine's Cosmography', *Augustinian Studies*, XXVII/2 (1996), pp. 129–77

Ferreiro, Larrie D., *Measure of the Earth* (New York, 2011)

Finkel, Irving, *The Ark before Noah: Decoding the Story of the Flood* (London, 2014)

Finocchiaro, Maurice A., *Retrying Galileo, 1633–1992* (Berkeley, CA, 2005)

Fox, Robin Lane, *Alexander the Great* (London, 2004)

Freedman, H., trans., 'Pesahim', in *The Babylonian Talmud*, ed.
   I. Epstein (London, 1938)
Frendo, David, 'Agathias' View of the Intellectual Attainments of
   Khusrau I: A Reconsideration of the Evidence', *Bulletin of the Asia
   Institute*, XVII (2004), pp. 97–110
Furley, David, 'Cosmology', in *The Cambridge History of Hellenistic
   Philosophy*, ed. Keimpe Algra et al. (Cambridge, 2005), pp. 412–51
Gainsford, Peter, 'The Eratosthenes Video Published by Business Insider:
   A Fact-Check', http://kiwihellenist.blogspot.com, 6 November
   2016
Garth, John, *The Worlds of J.R.R. Tolkien* (London, 2020)
Garwood, Christine, *Flat Earth: The History of an Infamous Idea*
   (London, 2007)
Geminos, *Geminos's Introduction to the Phenomena: A Translation and
   Study of a Hellenistic Survey of Astronomy*, ed. James Evans and J. L.
   Berggren (Princeton, NJ, 2006)
al-Ghazali, *The Confessions of Al-Ghazalli*, ed. Claude Field (London,
   1909)
—, *Incoherence of the Philosophers*, ed. Sabih Ahmad Kamali (Lahore,
   1963)
Gombrich, Richard F., 'Ancient Indian Cosmology', in *Ancient Cosmologies*,
   ed. Carmen Blacker and Michael Loewe (London, 1975), pp. 110–42
Goodman, Martin, *A History of Judaism* (London, 2019)
Gould, Stephen Jay, *Rocks of Ages: Science and Religion in the Fullness of
   Life* (London, 2001)
Graham, Daniel W., *Science before Socrates: Parmenides, Anaxagoras,
   and the New Astronomy* (Oxford, 2013)
Grant, Edward, ed., *A Source Book in Medieval Science* (Cambridge, MA,
   1974)
Green, Roger Lancelyn, and Walter Hooper, *C. S. Lewis: A Biography*
   (London, 2002)
Guan, Yuzhen, 'A New Interpretation of Shen Kuo's *Ying Biao Yi*',
   Archive for History of Exact Sciences, LXIV (2010), pp. 707–19
Gulácsi, Zsuzsanna, 'Matching the Three Fragments of the Chinese
   Manichaean Diagram of the Universe', *Studies on the Inner Asian
   Languages*, XXX (2015), pp. 79–94
—, and Jason BeDuhn, 'Picturing Mani's Cosmology: An Analysis
   of Doctrinal Iconography on a Manichaean Hanging Scroll from
   13th/14th-Century Southern China', *Bulletin of the Asia Institute*,
   XXV (2011), pp. 55–105
Gupta, R. C., 'Aryabhata', in *Encyclopaedia of the History of Science,
   Technology, and Medicine in Non-Western Cultures*, ed. Helaine Selin
   (Dordrecht, 1997), pp. 72–3
Gutas, Dimitri, *Greek Thought, Arabic Culture: The Graeco-Arabic
   Translation Movement in Baghdad and Early Abbāsid Society
   (2nd–4th/8th–10th Centuries)* (London, 1998)

Hannam, James, *God's Philosophers: How the Medieval World Laid the Foundations of Modern Science* (London, 2009)

Hansen, Valerie, *The Open Empire: A History of China to 1600* (New York, 2000)

Harlow, Daniel, 'Creation According to Genesis: Literary Genre, Cultural Context, Theological Truth', *Christian Scholar's Review*, XXXVII/2 (2008), pp. 163–98

Harrison, Peter, *The Territories of Science and Religion* (Chicago, IL, 2015)

Hayati, Said, 'Mar Aba I: Historical Context and Biographical Reconstruction', MA dissertation, University of Salzburg, 2018

Heath, Thomas L., *Aristarchus of Samos: The Ancient Copernicus* (Cambridge, 1913)

Heinen, Anton M., *Islamic Cosmology* (Beirut, 1982)

Henderson, John B., *The Development and Decline of Chinese Cosmology* (New York, 1984)

—, 'Ch'ing Scholars' Views of Western Astronomy', *Harvard Journal of Asiatic Studies*, XLVI/1 (1986), pp. 121–48

Henry, John, *Knowledge Is Power: How Magic, the Government and an Apocalyptic Vision Inspired Francis Bacon to Create Modern Science* (Cambridge, 2002)

Herodotus, *The Histories*, ed. Aubrey de Sélincourt and John M. Marincola (Harmondsworth, 1996)

Herren, Michael W., ed., *The Cosmography of Aethicus Ister: Edition, Translation, and Commentary* (Turnhout, 2011)

Hesiod, *Theogony*, ed. M. L. West (Oxford, 1966)

—, 'Theogony', in *Hesiod and Theognis*, ed. Dorothea Wender (Harmondsworth, 1973), pp. 23–58

—, 'Works and Days', in *Hesiod and Theognis*, ed. Dorothea Wender (Harmondsworth, 1973), pp. 59–86

Hewson, Robert, 'Science in Seventh-Century Armenia: Ananias of Sirak', *Isis*, LIX/1 (1968), pp. 32–45

Ho Peng Yoke, *Li, Qi and Shu: An Introduction to Science and Civilization in China* (Seattle, WA, 1987)

Holland, Tom, *In the Shadow of the Sword* (London, 2012)

Homer, *The Iliad*, ed. William F. Wyatt and A. T. Murray (Cambridge, MA, 1924)

Horace, 'Epistles', in *Satires. Epistles. The Art of Poetry*, ed. H. Rushton Fairclough (Cambridge, MA, 1926), pp. 248–441

Horky, Phillip Sidney, ed., *Cosmos in the Ancient World* (Cambridge, 2019)

Horowitz, Wayne, *Mesopotamian Cosmic Geography* (Winona Lake, IN, 1998)

Huff, Toby E., *Intellectual Curiosity and the Scientific Revolution: A Global Perspective* (Cambridge, 2011)

Huffman, Carl A., *Philolaus of Croton: Pythagorean and Presocratic. A Commentary on the Fragments and Testimonia with Interpretive Essays* (Cambridge, 1993)

Huntingford, G.W.B., ed., *The Periplus of the Erythraean Sea* (London, 1980)

Hutchings, David, and James C. Ungureanu, *Of Popes and Unicorns: Science, Christianity, and How the Conflict Thesis Fooled the World* (Oxford, 2021)

Huxley, George, 'Studies in the Greek Astronomers', *Greek, Roman, and Byzantine Studies*, IV/2 (1963), pp. 83–105

Ibn al-Nadīm, Muḥammad ibn Isḥāq, *The Fihrist of Al-Nadīm: A Tenth-Century Survey of Muslim Culture* (New York, 1970)

İhsanoğlu, Ekmeleddin, 'Modern Islam', in *Science and Religion around the World*, ed. John Hedley Brooke and Ronald L. Numbers (New York, 2011), pp. 148–74

Inglebert, Hervé, '"Inner" and "Outer" Knowledge: The Debate between Faith and Reason in Late Antiquity', in *A Companion to Byzantine Science*, ed. Stavros Lazaris (Leiden, 2020), pp. 27–52

Inwood, Brad, L. P. Gerson and D. S. Hutchinson, eds, *The Epicurus Reader: Selected Writings and Testimonia* (Indianapolis, IN, 1994)

Irving, Washington, *Life and Voyages of Christopher Columbus* (London, 1909)

Isidore of Seville, *The Etymologies of Isidore of Seville*, ed. Stephan A. Barney (Cambridge, 2006)

—, *On the Nature of Things*, ed. Calvin B. Kendall and Faith Wallis (Liverpool, 2016)

Janos, Damien, 'Qur'ānic Cosmography in Its Historical Perspective: Some Notes on the Formation of a Religious Worldview', *Religion*, XLII/2 (2012), pp. 215–31

Jefferson, Thomas, *Notes on the State of Virginia* (Boston, MA, 1832)

Jenkins, Philip, *The Lost History of Christianity* (New York, 2008)

John of Damascus, 'Exposition of the Orthodox Faith', in *St Hilary of Poitiers and John of Damascus*, ed. S.D.F. Salmond (Oxford, 1899)

Jones, Alexander, *Astronomical Papyri from Oxyrhynchus* (Philadelphia, PA, 1999)

—, *A Portable Cosmos: Revealing the Antikythera Mechanism, Scientific Wonder of the Ancient World* (Oxford, 2017)

Josephus, *Jewish Antiquities*, ed. H. St J. Thackeray (Cambridge, MA, 1930)

Julian, 'Fragment of a Letter to a Priest', in *Julian*, ed. Wilmer C. Wright (Cambridge, MA, 1913), vol. II, pp. 295–340

Kaldellis, Anthony, and Niketas Siniossoglou, *The Cambridge Intellectual History of Byzantium* (Cambridge, 2017)

Keas, Michael Newton, *Unbelievable: 7 Myths about the History and Future of Science and Religion* (Wilmington, DE, 2019)

Kenny, Anthony, *A New History of Western Philosophy* (Oxford, 2010)

Khatchadourian, Haig, Nicholas Rescher and Ya'qub ibn Ishaq
al-Kindi, 'Al-Kindi's Epistle on the Concentric Structure of the
Universe', *Isis*, LVI/2 (1965), pp. 190–95
King, David A., 'The Sacred Direction of Mecca: A Study of the
Interaction of Religion and Science in the Middle Ages',
Interdisciplinary Science Reviews, X/4 (1985), pp. 315–28
—, and Richard P. Lorch, 'Qibla Charts, Qibla Maps, and Related
Instruments', in *Cartography in Traditional Islamic and South Asian
Societies*, ed. David Woodward and J. B. Harley (Chicago, IL, 1992),
pp. 189–205
Kirk, G. S., J. E. Raven and M. Schofield, *The Presocratic Philosophers:
A Critical History with a Selection of Texts* (Cambridge, 1983)
Knibb, M. A., trans., '1 Enoch', in *The Apocryphal Old Testament*, ed.
H.F.D. Sparks (Oxford, 1984), pp. 169–320
Knudsen, Toke Lindegaard, *The Siddhāntasundara of Jñānarāja, an
English Translation with Commentary* (Baltimore, MD, 2014)
Koch-Westenholz, Ulla, 'Babylonian Views of Eclipses', *Res Orientalis*,
XIII (2001), pp. 71–84
Kominko, Maja, *The World of Kosmas* (Cambridge, 2013)
Kotyk, Jeffrey, 'Celestial Deities in the Flat-Earth Buddhist Cosmos
and Astrology', in *Intersections of Religion and Astronomy*, ed. Chris
Corbally, Darry Dinell and Aaron Ricker (London, 2020), pp. 36–43
—, 'The Chinese Buddhist Approach to Science: The Case of Astronomy
and Calendars', *Journal of Dharma Studies*, III/2 (2020), pp. 273–89
—, 'Examining Amoghavajra's Flat-Earth Cosmology: Religious vs
Scientific Worldviews in Buddhist Astrology', *Studies in Chinese
Religions*, VII/2 (2021), pp. 203–20
Krauss, Rolf, 'Egyptian Calendars and Astronomy', in *The Cambridge
History of Science*, vol. I: *Ancient Science*, ed. Alexander Jones and
Liba Taub (Cambridge, 2003), pp. 131–43
Lacey, Robert, *Inside the Kingdom* (London, 2009)
Lactantius, *The Divine Institutes*, ed. Anthony Bowen and Peter
Garnsey (Liverpool, 2003)
Laderman, Shulamit, *Images of Cosmology in Jewish and Byzantine Art*
(Leiden, 2013)
Laven, Mary, *Mission to China* (London, 2011)
Lehoux, Daryn, *What Did the Romans Know? An Inquiry into Science
and Worldmaking* (Chicago, IL, 2012)
Levathes, Louise, *When China Ruled the Seas: The Treasure Fleet of the
Dragon Throne, 1405–1433* (Oxford, 1994)
Lewis, C. S., *The Voyage of the Dawn Treader* [1952] (London, 1980)
—, *The Discarded Image: An Introduction to Medieval and Renaissance
Literature* [1964] (Cambridge, 1994)
Lindberg, David, 'Science and the Early Church', in *God and Nature:
Historical Essays on the Encounter between Science and Christianity*, ed.
David Lindberg and Ronald Numbers (Berkeley, CA, 1986), pp. 19–48

Liu An, *The Huainanzi: A Guide to the Theory and Practice of Government in Early Han China*, ed. John Major et al., trans. John Major (New York, 2010)

Livingstone, David N., *The Geographical Tradition: Episodes in the History of a Contested Enterprise* (Oxford, 1992)

Lloyd, G.E.R., *Magic, Reason and Experience: Studies in the Origins and Development of Greek Science* (Cambridge, 1979)

Lopez, Donald S., *Buddhism and Science: A Guide for the Perplexed* (Chicago, IL, 2008)

Losee, John, *A Historical Introduction to the Philosophy of Science* (Oxford, 2001)

Lucan, *The Civil War*, ed. J. D. Duff (Cambridge, MA, 1928)

Lucretius, *On the Nature of the Universe*, ed. R. E. Latham and John Godwin (Harmondsworth, 2005)

Macaulay, T. B., 'Macaulay's Minute', in *Selections from Educational Records, Part I (1781–1839)*, ed. H. Sharp (Delhi, 1965), pp. 107–17

McCluskey, Stephen C., *Astronomies and Cultures in Early Medieval Europe* (Cambridge, 1998)

McCready, William D., 'Isidore, the Antipodeans, and the Shape of the Earth', *Isis*, LXXXVII/1 (1996), pp. 108–27

McCrindle, J. W., ed., *The Christian Topography of Cosmas* (London, 1897)

Macrobius, *Commentary on the Dream of Scipio*, ed. William Harris Stahl (New York, 1952)

al-Maḥallī, Jalāl al-Dīn and Jalāl al-Dīn al-Suyūṭī, *Tafsīr Al-Jalālayn*, ed. Feras Hamza (Amman, 2007)

Maimonides, Moses, 'Foundation of the Torah', www.sefaria.org (1927)

—, *The Guide for the Perplexed*, ed. M. Friedländer (New York, 1956)

Mak, Bill M., 'Yusi Jing – a Treatise of "Western" Astral Science in Chinese and Its Versified Version *Xitian Yusi Jing*', SCIAMVS, XV (2014), pp. 105–39

Malter, Henry, *Saadia Gaon: His Life and Works* (Philadelphia, PA, 1921)

Mango, Cyril A., *Byzantium: The Empire of New Rome* (London, 1980)

Martianus Capella, *The Marriage of Philology and Mercury*, ed. William Harris Stahl and Richard Johnson (New York, 1977)

Martín, Inmaculada Pérez, and Gonzalo Cruz Andreotti, 'Geography', in *A Companion to Byzantine Science*, ed. Stavros Lazaris (Leiden, 2020), pp. 231–60

Mascaró, Juan, ed., *The Bhagavad Gita* (Harmondsworth, 1962)

Matt, Daniel Chanan, ed., *The Zohar* (Stanford, CA, 2004)

Meier, John P., *A Marginal Jew: The Roots of the Problem and the Person* (New York, 1991)

Miller, L., and Maurice Simon, 'Bekoroth', in *The Babylonian Talmud*, ed. I. Epstein (London, 1948)

Mimura, Taro, 'The Arabic Original of (Ps.) Māshā'Allāh's Liber De Orbe: Its Date and Authorship', *British Journal for the History of Science*, XLVIII/2 (2015), pp. 321–52

Minkowski, Christopher, 'Competing Cosmologies in Early Modern Indian Astronomy', in *Ketuprakāśa: Studies in the History of the Exact Sciences in Honor of David Pingree*, ed. Charles Burnett, Jan Hogendijk and Kim Plofker (Leiden, 2004), pp. 349–85

Morrissey, Fitzroy, *A Short History of Islamic Thought* (London, 2021)

Moseley, C.W.R.D., ed., *The Travels of Sir John Mandeville* (Harmondsworth, 1983)

Natali, Carlo, *Aristotle* (Princeton, NJ, 2013)

Needham, Joseph, and Ling Wang, *Mathematics and the Sciences of the Heavens and the Earth* (Cambridge, 1959)

Netz, Reviel, 'The Bibliosphere of Ancient Science (outside of Alexandria): A Preliminary Survey', *Naturwissenschaften, Technik und Medizin*, XIX/3 (2011), pp. 239–69

Neugebauer, Otto, *The Exact Sciences in Antiquity* (New York, 1969)

Newton, Isaac, 'Principia', in *On the Shoulders of Giants: The Great Works of Physics and Astronomy*, ed. Stephen Hawking (Philadelphia, PA, 2002), pp. 733–1160

Nothaft, C.P.E., 'Augustine and the Shape of the Earth: A Critique of Leo Ferrari', *Augustinian Studies*, XLII/1 (2011), pp. 33–48

O'Flaherty, Wendy, ed., *Hindu Myths* (Harmondsworth, 1975)

——, *The Rig Veda* (Harmondsworth, 1981)

Obbink, Dirk, 'Lucretius and the Herculaneum Library', in *The Cambridge Companion to Lucretius*, ed. Stuart Gillespie and Philip R. Hardie (Cambridge, 2007), pp. 33–40

Obrist, Barbara, 'William of Conches, Māshā'Allāh, and Twelfth-Century Cosmology', *Archives d'Histoire Doctrinale et Littéraire du Moyen Âge*, LXXVI/1 (2009), pp. 29–87

Ovid, *Metamorphoses*, ed. Mary Innes (Harmondsworth, 1955)

Paine, Thomas, 'The Age of Reason Part One', in *The Thomas Paine Reader*, ed. Michael Foot (Harmondsworth, 1987), pp. 399–451

Panchenko, Dimitri, 'Anaxagoras' Argument against the Sphericity of the Earth', *Hyperboreus*, III/1 (1997), pp. 175–8

Parry, J. H., *The Age of Reconnaissance: Discovery, Exploration and Settlement 1450 to 1650* (London, 1963)

Philo, 'A Treatise on the Cherubim', in *Philo*, ed. F. H. Colson and G. H. Whitaker (Cambridge, MA, 1929), pp. 3–87

——, *Questions of Genesis*, ed. Ralph Marcus (Cambridge, MA, 1953)

Philoponus, John, *De Opificio Mundi*, ed. Clemens Scholten (Freiburg, 1997)

Photius, *The Bibliotheca*, ed. Nigel Wilson (London, 1994)

Pingree, David, 'The Fragments of the Works of Al-Fazārī', *Journal of Near Eastern Studies*, XXIX/2 (1970), pp. 103–23

——, *The Yavanajātaka of Sphujidhvaja* (Cambridge, MA, 1978)

——, 'The Purāṇas and Jyotiḥśāstra: Astronomy', *Journal of the American Oriental Society*, CX/2 (1990), pp. 274–80

Pingree, David, and C. J. Brunner, 'Astronomy and Astrology in
Iran', in *Encyclopædia Iranica* (London, 1987), vol. 11/8,
pp. 858–71
Plato, 'Cratylus', in *Plato: Cratylus and Others*, ed. Harold North
Fowler (Cambridge, MA, 1926), pp. 1–192
—, 'Parmenides', in *Plato: Cratylus and Others*, ed. Harold North
Fowler (Cambridge, MA, 1926), pp. 193–332
—, 'Phaedrus', in *Phaedrus and Letters VII and VIII*, ed. Walter
Hamilton (Harmondsworth, 1973), pp. 19–103
—, *The Laws*, ed. Trevor J. Saunders (Harmondsworth, 1975)
—, 'Timaeus', in *Timaeus and Critias*, ed. Desmond Lee
(Harmondsworth, 1977), pp. 27–126
—, *Theaetetus*, ed. Robin A. H. Waterfield (Harmondsworth, 1987)
—, 'Phaedo', in *The Last Days of Socrates*, ed. Hugh Tredennick
and Harold Tarrant (Harmondsworth, 1993), pp. 93–185
Pliny the Elder, *Natural History*, ed. H. Rackham (Cambridge, MA,
1938)
Pliny the Younger, *The Letters of the Younger Pliny*, ed. Betty Radice
(Harmondsworth, 1969)
Plofker, Kim, 'Astronomy and Astrology on India', in *The Cambridge
History of Science*, vol. 1: *Ancient Science*, ed. Alexander Jones and
Liba Taub (Cambridge, 2003), pp. 485–500
—, 'Derivation and Revelation: The Legitimacy of Mathematical
Models in Indian Cosmology', in *Mathematics and the Divine: A
Historical Study*, ed. T. Koetsier and L. Bergmans (Amsterdam,
2004), pp. 61–76
—, *Mathematics in India* (Princeton, NJ, 2009)
Plumley, J. M., 'The Cosmology of Ancient Egypt', in *Ancient
Cosmologies*, ed. Carmen Blacker and Michael Loewe (London,
1975), pp. 17–41
Plutarch, 'Concerning the Face Which Appears in the Orb of the
Moon', in *Moralia*, ed. Harold Cherniss and W. C. Helmbold
(Cambridge, MA, 1957), vol. XII, pp. 34–226
—, 'Life of Pericles', in *The Rise and Fall of Athens: Nine Greek Lives*, ed.
Ian Scott-Kilvert (Harmondsworth, 1960), pp. 165–206
Popper, Karl, *Conjectures and Refutations: The Growth of Scientific
Knowledge* (London, 2002)
Portuondo, Maria M., 'Lunar Eclipses, Longitude and the New World',
Journal of the History of Astronomy, XL/3 (1990), pp. 249–76
Pratchett, Terry, *The Colour of Magic* [1983] (London, 1985)
Priscian, *Answers to King Khosroes of Persia*, ed. Pamela Huby (London,
2016)
Pritchard, James B., *Ancient Near Eastern Texts Relating to the Old
Testament* (Princeton, NJ, 1969)
Pseudo-Plato, 'Epinomis', in *Plato: Charmides et al.*, ed. W.R.M. Lamb
(Cambridge, MA, 1927), pp. 423–87

Pseudo-Xenophon, 'The Constitution of Athens', in *Scripta Minora*, ed. E. C. Marchant and G. W. Bowersock (Cambridge, MA, 1925), pp. 459–508

Ptolemy of Alexandria, *Ptolemy's Almagest*, ed. G. J. Toomer (London, 1984)

—, *Ptolemy's Geography: An Annotated Translation of the Theoretical Chapters*, ed. J. L. Berggren and Alexander Jones (Princeton, NJ, 2000)

Quintus Curtius Rufus, *The History of Alexander*, ed. John Yardley and Waldemar Heckel (Harmondsworth, 1984)

Rabbi Eliezar, *Pirke De Rabbi Eliezer*, ed. Gerald Friedlander (London, 1916)

Ramírez-Weaver, Eric M., 'William of Conches, Philosophical Continuous Narration, and the Limited Worlds of Medieval Diagrams', *Studies in Iconography*, XXX (2009), pp. 1–41

Randles, W.G.L., 'The Evaluation of Columbus' "India" Project by Portuguese and Spanish Cosmographers in the Light of the Geographical Science of the Period', *Imago Mundi*, XLII (1990), pp. 50–64

Rateliff, John, 'The Flat Earth Made Round and Tolkien's Failure to Finish the Silmarillion', *Journal of Tolkien Research*, IX/1 (2020), pp. 1–17

Rochberg, Francesca, *The Heavenly Writing: Divination, Horoscopy, and Astronomy in Mesopotamian Culture* (Cambridge, 2004)

—, 'Babylonian Astral Science in the Hellenistic World: Reception and Transmission', CAS ESERIES, IV (2010), pp. 1–11

—, *Before Nature: Cuneiform Knowledge and the History of Science* (Chicago, IL, 2016)

Rochberg-Halton, Francesca, 'Babylonian Horoscopes and Their Sources', *Orientalia*, LVIII/1 (1989), pp. 102–23

Romm, James S., *The Edges of the Earth in Ancient Thought: Geography, Exploration, and Fiction* (Princeton, NJ, 1992)

Roux, Georges, *Ancient Iraq* (London, 1992)

Rowbotham, Samuel Birley ("Parallax"), *Zetetic Astronomy: Earth Not a Globe* (London, 1881)

Russell, Bertrand, *History of Western Philosophy and Its Connection with Political and Social Circumstances from the Earliest Times to the Present Day* (London, 1961)

Russell, Jeffrey Burton, *Inventing the Flat Earth: Columbus and Modern Historians* (Westport, VA, 1991)

Saadia ben Joseph, *Commentaire sur Le Séfer Yesira: Ou Livre de la Création* (Paris, 1891)

—, *Rabbi Saadiah Gaon's Commentary on the Book of Creation*, ed. Michael Linetsky (Northvale, NJ, 2002)

Sagan, Carl, *Pale Blue Dot* (London, 1994)

Saliba, George, *Islamic Science and the Making of the European Renaissance* (Cambridge, MA, 2007)

Sambursky, Samuel, *Physics of the Stoics* (Princeton, NJ, 1959)

Sanders, N. K., ed., 'The Babylonian Creation', in *Poems of Heaven and Hell from Ancient Mesopotamia*, ed. N. K. Sanders (Harmondsworth, 1971), pp. 11–112

——, *The Epic of Gilgamesh* (Harmondsworth, 1972)

——, *Encyclopaedia of the History of Science, Technology, and Medicine in Non-Western Cultures* (Dordrecht, 1997)

Scafi, Alessandro, 'Defining Mappamundi', in *The Hereford World Map: Medieval World Maps and Their Context*, ed. P.D.A. Harvey (London, 2006), pp. 345–54

Sedley, D. N., *Lucretius and the Transformation of Greek Wisdom* (Cambridge, 1998)

Selin, Helaine, ed., *Encyclopaedia of the History of Science, Technology, and Medicine in Non-Western Cultures* (Dordrecht, 1997)

——, ed., *Astronomy Across Cultures: A History of Non-Western Astronomy* (Dordrecht, 2000)

Seneca the Younger, 'Medea', in *Tragedies*, ed. John G. Fitch (Cambridge, MA, 2018), vol. I, pp. 305–406

Severus Sebokht, 'Description of the Astrolabe', in *Astrolabes of the World*, ed. R. T. Gunther (Oxford, 1932), pp. 82–103

Sharma, R. S., *India's Ancient Past* (New Delhi, 2005)

Sidrys, Raymond V., *The Mysterious Spheres on Greek and Roman Ancient Coins* (Oxford, 2020)

Silva, Marco, 'Flat Earth: How Did YouTube Help Spread a Conspiracy Theory?', www.bbc.co.uk, 18 July 2019

Simon-Shoshan, Moshe, 'The Heavens Proclaim the Glory of God: A Study in Rabbinic Cosmology', *Bekhol Derakhekha Daehu – Journal of Torah and Scholarship*, XX (2008), pp. 67–96

Skjærvø, Prods Oktor, *The Spirit of Zoroastrianism* (New Haven, CT, 2011)

Slifkin, Natan, *The Sun's Path at Night: The Revolution in Rabbinic Perspective on the Ptolemaic Revolution*, www.zootorah.com (2012)

Smyth, Marina, *Understanding the Universe in Seventh-Century Ireland* (Woodbridge, 1996)

——, 'The Seventh-Century Hiberno-Latin Treatise "Liber De Ordine Creaturarum", a Translation', *Journal of Medieval Latin*, XXI (2011), pp. 137–222

Sorabji, Richard, 'John Philoponus', in *Philoponus and the Rejection of Aristotelian Science*, ed. Richard Sorabji (London, 2010)

Starr, S. Frederick, *Lost Enlightenment: Central Asia's Golden Age from the Arab Conquest to Tamerlane* (Princeton, NJ, 2013)

Sterckx, Roel, *Chinese Thought: From Confucius to Cook Ding* (London, 2019)

Stevens, W. M., 'The Figure of the Earth in Isidore's "De Natura Rerum"', *Isis*, LXXI/2 (1980), pp. 268–77

Strabo, *Geography*, ed. Horace Leonard Jones (Cambridge, MA, 1930)

Subbarayappa, Bidare V., and K. V. Sarma, eds, *Indian Astronomy: A Source-Book* (Bombay, 1985)
Suetonius, 'On Grammarians', in *Suetonius*, ed. J. C. Rolfe (Cambridge, MA, 1914), vol. II, pp. 378–417
al-Tabari, Muhammad ibn Yarir, *The History of Al-Tabari* (New York, 1989)
Tabataba'i, Mohammad Ali and Saida Mirsadri, 'The Qur'ānic Cosmology, as an Identity in Itself', *Arabica*, LXIII/3–4 (2016), pp. 201–34
Tattersall, Jill, 'Sphere or Disc? Allusions to the Shape of the Earth in Some Twelfth-Century and Thirteenth-Century Vernacular French Works', *Modern Language Review*, LXXVI/1 (1981), pp. 31–46
Thackeray, H. St J., *The Letter to Aristeas: Translated with an Appendix of Ancient Evidence of the Origin of the Septuagint* (London, 1917)
Thapar, Romila, *Early India: From the Origins to AD 1300* (Berkeley, CA, 2002)
Thucydides, *History of the Peloponnesian War*, ed. Rex Warner and M. I. Finley (Harmondsworth, 1972)
Tibbetts, Gerald R., 'The Beginnings of a Cartographic Tradition', in *Cartography in Traditional Islamic and South Asian Societies*, ed. David Woodward and J. B. Harley (Chicago, IL, 1992), pp. 90–107
Tihon, Anne, 'Astronomy', in *The Cambridge Intellectual History of Byzantium*, ed. Anthony Kaldellis and Niketas Siniossoglou (Cambridge, 2017), pp. 183–97
Tolkien, J.R.R., *The Lost Road and Other Writings* (London, 1987)
—, *The Monsters and the Critics and Other Essays* (London, 1997)
—, *Beowulf: A Translation and Commentary, Together with Sellic Spell* (London, 2014)
—, *Tolkien on Fairy-Stories*, ed. Verlyn Flieger and Douglas A. Anderson (London, 2014)
Trigger, Bruce, *Understanding Early Civilisations: A Comparative Study* (Cambridge, 2003)
Tyldesley, Joyce, *The Penguin Book of Myths and Legends of Ancient Egypt* (London, 2011)
Tylor, Edward Burnett, *Anthropology* (London, 1913)
Van Bladel, Kevin, 'Heavenly Cords and Prophetic Authority in the Quran and Its Late Antique Context', *Bulletin of the School of Oriental and African Studies* (2007), pp. 223–46
—, 'The Arabic History of Science of Abū Sahl ibn Nawbaḫt (fl. ca 770–809) and Its Middle Persian Sources', in *Islamic Philosophy, Science, Culture, and Religion*, ed. Felicitas Opwis and David Reisman (Leiden, 2012), pp. 41–62
—, 'Eighth-Century Indian Astronomy in the Two Cities of Peace', in *Islamic Cultures, Islamic Contexts: Essays in Honor of Professor Patricia Crone*, ed. Behnam Sadeghi et al. (Leiden, 2014), pp. 257–9

Vernant, Jean-Pierre, *The Origins of Greek Thought* (Ithaca, NY, 1982)
Virgil, 'Georgics', in *Eclogues. Georgics. Aeneid: Books 1–6*, ed.
H. Rushton Fairclough (Cambridge, MA, 1916), pp. 79–237
Vrhovski, Jan, 'Apologeticism in Chinese Nestorian Documents from
the Tang Dynasty', *Asian Studies* I, XVII/2 (2013), pp. 53–70
Walton, John H., *Ancient Near Eastern Thought and the Old Testament:
Introducing the Conceptual World of the Hebrew Bible* (Grand Rapids,
MI, 2018)
Warren, James, 'Lucretius and Greek Philosophy', in *The Cambridge
Companion to Lucretius*, ed. Stuart Gillespie and Philip R. Hardie
(Cambridge, 2007), pp. 19–32
West, M. L., *The Hymns of Zoroaster* (London, 2010)
White, Andrew Dickson, *A History of the Warfare of Science with
Theology in Christendom* (New York, 1896)
Whitehead, Alfred North, *Process and Reality: An Essay in Cosmology*
(New York, 1978)
Wilkinson, Toby, *The Rise and Fall of Ancient Egypt: The History of a
Civilisation from 3000 BC to Cleopatra* (London, 2010)
William of Conches, *A Dialogue on Natural Philosophy: (Dragmaticon
Philosophiae)*, ed. Italo Ronca and Matthew Curr (Notre Dame, IN,
1997)
Williams, J., 'Isidore, Orosius and the Beatus Map', *Imago Mundi*, XLIX
(1997), pp. 7–32
Wilson, David B., 'The Historiography of Science and Religion', in
Science and Religion: A Historical Introduction, ed. Gary B. Ferngren
(Baltimore, MD, 2002), pp. 13–29
Wilson, H. H., *The Vishńu Puráńa: A System of Hindu Mythology and
Tradition* (London, 1840)
Woodward, David, 'Medieval *Mappaemundi*', in *Cartography in
Prehistoric, Ancient, and Medieval Europe and the Mediterranean*,
ed. David Woodward and J. B. Harley (Chicago, IL, 1987),
pp. 286–370
Wootton, David, *The Invention of Science: A New History of the Scientific
Revolution* (London, 2015)
Wright, J. Edward, *The Early History of Heaven* (New York, 2000)
Xenophon, 'Memoires of Socrates', in *Conversations of Socrates*, ed.
Robin Waterfield and Hugh Tredennick (Harmondsworth, 1990),
pp. 51–216
Yang, Qiao, 'From the West to the East, from the Sky to the Earth: A
Biography of Jamāl al-Dīn', *Asiatische Studien – Études Asiatiques*,
LXXI/4 (2018), pp. 1231–45
Zhang, Qiong, *Making the New World Their Own: Chinese Encounters
with Jesuit Science in the Age of Discovery* (Leiden, 2015)
Zhenghai, Song, and Chen Chuankang, 'Why Did Zheng He's Sea
Voyage Fail to Lead the Chinese to Make the "Great Geographic
Discovery"?', in *Chinese Studies in the History and Philosophy of*

Science and Technology: Boston Studies in the Philosophy of Science,
ed. Fan Dainian and Robert S. Cohen (Dordrecht, 1996), pp. 303–14

Zumeng, Jin, 'A Critique of "Zhang Heng's Theory of a Spherical
Earth"', in *Chinese Studies in the History and Philosophy of Science
and Technology,* ed. Fan Dainian and Robert S. Cohen (Dordrecht,
1996), pp. 427–32

ACKNOWLEDGEMENTS

This book covers the shape of the Earth in cultures all over the world. Like Shakespeare, I have little Latin and less Greek, so I am tremendously grateful to the scholars, both living and long dead, who translated the texts I have used into English from Akkadian, Arabic, Chinese, Hebrew, Persian, Sanskrit and other languages. Also, since the References do not always adequately reflect the inspiration that I have received from some of the scholarship I have read in the course of researching this book, I would like to record my gratitude to the following, even though I have not always agreed with them: Daniel W. Graham on the Presocratics, the late D. R. Dick on Greek science, Kim Plofker on Indian astronomy, Martin Goodman on Judaism, Donald S. Lopez on Buddhism, Jonathan A. C. Brown on Islam, Felipe Fernández-Armesto on exploration, Christopher Cullen on Chinese cosmology and Christine Garwood on modern flat-earthers.

Many people kindly gave their time to look at draft chapters or even the whole book, making valuable comments and correcting some howlers. They include Pieter Beullens, Thony Christie, Peter Gainsford, Boaz Goran, Laura Hassan, Jeffrey Kotyk, Reginald O'Donoghue, Tim O'Neill, Khodadad Rezakhani, Jim Slagle and James Ungureanu. I am entirely responsible for errors that remain and for the opinions expressed in these pages. My immensely talented daughter Alexandra produced the diagrams of the various world pictures, clarifying my attempts to describe them in words.

I was thrilled when Dave Watkins commissioned this book for Reaktion. I am also very grateful to Michael Leaman, Amy Salter and Alex Ciobanu for all their work seeing it through the press, and to Francis Young for the index.

Finally, I want to thank the staff of the London Library, who were lifesavers during lockdown, posting out books as I requested them and then letting me hang on to them for ages.

PHOTO ACKNOWLEDGEMENTS

The author and publishers wish to express their thanks to the below sources of illustrative material and/or permission to reproduce it. Some locations of artworks are also given below, in the interest of brevity:

akg-images: p. 168 (Trongsa Dzong); Alamy Stock Photo: pp. 20 (Prisma by Dukas Presseagentur GmbH), 35 (Granger Historical Picture Archive), 252 (View Stock); from Peter Apian, *Cosmographia* (Antwerp, 1550), photo courtesy of the Linda Hall Library of Science, Engineering and Technology, Kansas City, MO (CC BY 4.0): p. 90; Bayerische Staatsbibliothek, Munich (Clm 14300, fol. 6v): p. 209; Biblioteca Medicea Laurenziana, Florence (MS Pluteus IX.28, fol. 95v): p. 174; Biblioteca Nacional de España, Madrid (Res 28, fol. 49r): p. 165 (*top*); Biblioteca Nazionale Marciana, Venice: p. 166 (*top*); Biblioteka Narodowa, Warsaw: pp. 162–3; photo Bonhams: p. 234 (Musée du Louvre, Paris); The British Museum, London: p. 26; Chiesa del Gesù, Rome: p. 268; from Christopher Clavius, *In sphaeram Ioannis de Sacro Bosco commentarius* (Rome, 1581), photo Wellcome Collection, London: p. 221; Flickr.com: p. 15 (photo Carole Raddato, CC BY-SA 2.0 – The British Museum, London); Germanisches Nationalmuseum, Nuremberg (CC BY-SA 4.0): p. 238; Hallwylska museet, Stockholm, photo Jenny Bergensten: p. 38 (Museo Archeologico Nazionale, Naples); courtesy of Alexandra Hannam: pp. 33 (after Mary Boyce, ed. and trans., *Textual Sources for the Study of Zoroastrianism* (Manchester, 1984)), 71 (after M. A. Orr (Evershed), *Dante and the Early Astronomers* (London and Edinburgh, 1913)), 109, 276 (after Xiong Mingyu, *Gezhi Cao* 格致草 (1648)); Hereford Cathedral: p. 230; photos courtesy of Chris Johnson: p. 114; The J. Paul Getty Museum, Los Angeles (MS Ludwig XV 4, fol. 156v): p. 111; Library of Congress, Washington, DC: pp. 167, 229; The Metropolitan Museum of Art, New York: pp. 29, 222, 251; Museo Nacional del Prado, Madrid: p. 242; Nanjing Museum: p. 166 (*bottom*); NASA: pp. 161, 296; The New York Public Library (MA 69, fol. 113v): p. 59; private collection: p. 164;

INDEX